IN SEARCH OF THE

Ivory-Billed
WOODPECKER

IN SEARCH OF THE

Ivory-Billed
WOODPECKER

Jerome A. Jackson

Smithsonian Books
Washington

Dedicated to James Taylor Tanner, ecologist, ornithologist, and friend—
for the window he gave us on the world of the
ivory-billed woodpecker;

to my parents, who encouraged me but often wondered
just what I was doing;

and to my wife Bette and my children, who encouraged,
helped, and endured.

© 2004 by Smithsonian Institution
All rights reserved

Copy editor: Debbie Hardin
Production editor: Robert A. Poarch
Designer: Brian Barth

Library of Congress Cataloging-in-Publication Data
Jackson, Jerome A.
  In search of the ivory-billed woodpecker / Jerome A. Jackson.
      p.   cm.
  Includes bibliographical references (p.).
  ISBN 1-58834-132-1 (alk. paper)
  I. Ivory-billed woodpecker. I. Title.
  QL696.P5J24 2004
  598.7'2—dc22   2004040944

British Library Cataloguing-in-Publication Data available

Manufactured in the United States of America
10 09 08 07 06 05 04   1 2 3 4 5

∞ The paper used in this publication meets the minimum requirements of the American
National Standard for Information Sciences—Permanence of Paper for Printed Library
Materials ANSI Z39.48-1984.

For permission to reproduce illustrations appearing in this book, please correspond
directly with the owners of the works, as listed in the individual captions. Smithsonian
Books does not retain reproduction rights for these illustrations individually or maintain a
file of addresses for photo sources.

Contents

Acknowledgments

$\mathcal{I}$ am indebted to many people who provided assistance in the field, guidance in various ways, reviews of portions of the manuscript, helpful references, and stimulating discussion. Among these I especially thank Jim and Nancy Tanner–Jim for his early encouragement, friendship, and the wealth of information that he shared–Nancy for her own observations and for making Jim's notes and other records and photos available to me. I also thank their son David for help with biographical material for his father.

I especially thank Richard N. Conner, William E. "Ted" Davis Jr., and my wife Bette Jackson, who reviewed an early draft of the manuscript, encouraged me along the way, and at times joined me in the field, and my son Jerry Jackson Jr., who took, edited, or restored many of the photos I've used in the book. Lester Short was responsible for getting me into ivory-billed woodpecker habitat in Cuba on my first trip, when I accompanied him, Jennifer Horne, Christophe Imboden, and

Montserrat Carbonell to assist with evaluation of the status of conservation in Cuba for the International Council for Bird Preservation. Les has also generously shared his vast knowledge of woodpeckers with me on numerous occasions. Orlando Garrido, Giram Gonzalez, Arturo Kirkconnell, Giraldo Alayón, Pedro Rosabal, and many other Cuban colleagues generously shared their hospitality, friendship, and scientific expertise during my various efforts in Cuba. I treasure the time spent in the field with each of them. My third trip to Cuba, in the spring of 1988, was funded and organized by the National Geographic Society. In addition to my Cuban colleagues, Bates Littlehales, Ted Parker, Bill Curtsinger, and Rob Hernandez shared that trip with me. My 2002 visit to the Sierra Maestra was made possible by Park Wright and my Cuban hosts. My work in the United States between 1986 and 1988 was funded in part by the US Fish and Wildlife Service. Several Fish and Wildlife Service professionals provided essential help with funding, logistics, and old records. I especially thank James Lewis. Others who shared information about ivory-bills and helped in many different ways include David Allen, Richard Banks, Laurie Binford, John Dennis, Mary Gustafson, John William Hardy, Jamie Hill, Malcolm Hodges, Danny Ingold, Douglas James, Lloyd Kiff, John Kricher, Roxie Laybourne, George Lowery, Peter McBride, John McNeely, William Bird Mounsey, Oscar Owre, Alan Poole, George Reynard, Kate Somerville, Susan Stans, Phillips B. Street, George Miksch Sutton, Paul Sykes, Ann Tarbell, James Tate Jr., and George Watson.

I am very grateful to the following museum curators, archivists, libraries, and museums for assistance with specimens, illustrations, and archival materials in their care: Stephen H. Amos and Charles C. Browne, Fairbanks Museum; Anniston [Alabama] Museum of Natural History; Luis Baptista, California Academy of Sciences; Jon C. Barlow, Royal Ontario Museum; William H. Barnard, Norwich University; Walter Breckenridge, Scott Lanyon, Robert Zink, and James Ford Bell,

Museum of Natural History; Ron C. Brister and Margaret Witt, Memphis Pink Palace Museum; Robert Dickerman; Kimball Garrett, Los Angeles County Museum of Natural History; Harlo Hadow, Coe College; Gene K. Hess, Delaware Museum of Natural History; Janet Hinshaw, Van Tyne Library, University of Michigan Museum of Zoology; Kathleen Jacklin, Cornell University Libraries; Ned Johnson, University of California, Berkeley; Richard Johnston and Robert Mengel, University of Kansas Museum of Natural History; Robert S. Kennedy, Cincinnati Museum of Natural History; Lloyd Kiff, Western Foundation of Vertebrate Zoology; Daniel Klem, Muhlenberg College; Thomas W. Little, Burpee Museum of Natural History, Rockford, Illinois; James Northern, University of California, Los Angeles; F. Scott Orcutt, University of Akron; Henri Ouellet, National Museums of Canada; Kenneth C. Parkes, Carnegie Museum of Natural History; Ray Paynter, Museum of Comparative Zoology, Harvard University; Roger Tory Peterson; Will Post, Charleston Museum; Charles Preston, Denver Museum of Natural History; J. Van Remsen, Steve Cardiff, and George Lowery, Museum of Natural Science, Louisiana State University; Peter Lowther, Field Museum of Natural History; George Schrimper, University of Iowa; Lester Short, Wes Lanyon, and Peter Capainolo, American Museum of Natural History; Fred Sibley, Yale University; Alfred E. Smalley, Tulane University; Robert Storer, University of Michigan Museum of Zoology; Scott Sutcliffe, Cornell Laboratory of Ornithology; Dawn Sher Thomae and Ann McMullen, Milwaukee Public Museum; Tom Webber, Florida Museum of Natural History; Richard Weisbrod, Cornell University; Glen Woolfenden, Archbold Biological Station; and V. Zoskot, Zoological Institute, Russian Academy of Sciences.

Without doubt there are dozens more names that should be added to this list of individuals and institutions that shared information, specimens, and archival materials, and hundreds who reported observations, shared memories, and encouraged my efforts. My heartfelt thanks to all.

Finally, I thank all of those associated with Smithsonian Books who facilitated publication of this book–especially Peter Cannell who encouraged me to write it, and Nicole Sloan, Robert Poarch, and Debbie Hardin, who saw it through the editorial process.

Introduction

The Feathered Grail

I began dreaming of finding an ivory-billed woodpecker (*Campephilus principalis*) during the spring of 1965, while I was a last-semester undergraduate at Iowa State University. That was the semester I had enrolled in ornithology and, as a consequence, became fascinated by the unique adaptations and interesting behaviors of woodpeckers. It did not escape my attention that the ivory-billed woodpecker, perhaps gone forever, had become the Holy Grail of birders—the one sighting that every birder fantasized about. I looked for ivory-bills when I went birding in Shimek State Forest in southeast Iowa, and when I birded in the nearby bottomlands along the Mississippi and Skunk Rivers. It made little difference to me that there was not even a single record of ivory-billed woodpeckers having occurred in Iowa. I was enthralled with the size of some of the oaks (*Quercus* sp.) and cottonwoods (*Populus deltoides*) in the bottomlands and with the diversity and abundance of woodpeckers. Each glimpse of a pileated woodpecker (*Dryocopus pileatus*) caused a momentary racing of my heart. Yes, I thought, it was possible. They could have just moved up the river, expanding their range as their southern habitats disappeared, could they not?

Perhaps I should back up a bit. Just what is an ivory-bill? And why had it become the Holy Grail of birding? Ivory-billed woodpeckers are the largest woodpeckers known to inhabit the United States, about twenty inches long, weighing about $1^1/_4$ pounds, and with a thirty-one-inch wingspan. It is no wonder that early settlers in the southeast referred to them as the "Lord God Woodpecker." I can easily imagine that on seeing such a woodpecker for the first time one might silently exhort, "Lord God, what a woodpecker!"

"America's largest woodpecker": that alone made it worth the effort to see. But seeing an ivory-bill is about more than size. The ivory-billed woodpecker is also a spectacularly patterned bird. It is the avian equivalent of a tuxedoed aristocrat. In April 1935, famed ornithlogist/artist George Miksch Sutton accompanied a Cornell University expedition to virgin forest along the Tensas River in northeast Louisiana to study ivory-billed woodpeckers. There he was able to sketch ivory-bills as they brought food to their young and interacted with one another at the nest site. He later painted these magnificent birds at their nest, bathed in the glory of early morning sun (Figure I.1).

Size and beauty seem compelling reasons for the search, but sightings are also coveted because we know only fragments about the bird's behavior. Every verified sighting of an ivory-bill has the potential to provide new clues about the woodpecker's behavior and ecology. Any documented observation, no matter how minor, would be welcomed for publication by well-respected ornithological journals.

The last reason, and probably the strongest motivator, is the rarity of the ivory-bill. Perhaps always relatively uncommon, by 1900 the ivory-billed woodpecker had become the rarest woodpecker in North America, perhaps in the world.

Thus, in 1965 I counted myself among the elite birders and professional ornithologists who thought themselves capable of seeing the elusive woodpecker. One glimpse was all I sought, but for me 1965 ended without suc-

Figure I.1. Painting by ornithologist/artist George Miksch Sutton of a pair of ivory-billed woodpeckers at their nest in the Singer Tract, an area of virgin forest along the Tensas River in northeast Louisiana. This is the same nest and pair shown in the photo by Arthur Allen in Figure 1.1. *Courtesy George Miksch Sutton*

cess. A year later, in the fall of 1966, during my first semester as a graduate student, I saw—and held—my first ivory-bill specimen. I gingerly laid it out side by side with a stuffed pileated woodpecker on top of a specimen case at the University of Kansas Museum of Natural History. By itself, the pileated was impressive; next to the ivory-bill, it was puny. It was not that the body of the ivory-bill was so much larger than that of the pileated, but rather that the bill of the ivory-bill was so much larger and so different. Not just lighter in color, but heavier, flattened at the tip like a carpenter's wood chisel. Its feet also seemed bigger and its claws longer and more curved, its crest longer and more pointed, its tail stiffer and with the vanes of feathers uniquely curved inward to form somewhat of a trough on the underside. I not only wanted to see the ivory-bill, I wanted to study its behavior and ecology, to understand its adaptations for existence, to understand and be able to do something about its precarious status.

That autumn I read Robert K. Selander's seminal article *Sexual Dimorphism and Differential Niche Utilization in Woodpeckers*, a study that demonstrated that not only did different woodpeckers feed in different places and on different foods but that within some species of woodpeckers, males and females had bills that differed in size, reflecting further subdivision of the species niche.[1] Selander suggested that as a result of such differences the sexes might compete less for food and thus might have a stronger pair bond. I became hooked on studying how woodpeckers shared resources within and among species. I measured my first ivory-bills at the University of Kansas, wondering how males and females differed in size and how they might behave differently. As I traveled from museum to museum, examining specimens of downy (*Picoides pubescens*) and hairy woodpeckers (*Picoides villosus*) for my dissertation research, I also examined every ivory-bill specimen I could get my hands on. I was determined that someday I would see one flying.

In 1970, I was offered and accepted a position at Mississippi State University. It was more than my first academic position; I was moving to ivory-bill country. I quickly found that there were specimen records from within twenty miles of the Mississippi State University campus. Ivory-bills went from dream to possibility.

For thirty years as I traveled about the South in the course of my teaching and research, I kept a search image of ivory-bills in my mind. My first copy of James Tanner's 1942 monograph on the great woodpecker became tattered as a result of constant use.[2] From South Carolina's Santee River, to Georgia's Okefenokee Swamp and Altamaha River, to Florida's Apalachicola and Chipola Rivers and the Fakahatchee Strand, to the Big Thicket of Texas, the Atchafalaya Basin and Pearl River Swamp of Louisiana, and to Tibbee Creek and the lower Yazoo River in Mississippi, whenever I went into suitable habitat, I thought of ivory-bills.

Many of the places that were once home to ivory-bills had felt the long reach of human progress, and the great birds were gone. Yet remnants of the primordial southern river forests remained. I found a copy of Bernard Romans's *A Natural History of East and West Florida* in which he described canoeing down the sinuous Tombigbee River in 1771.[3] His descriptions were so vivid that when I canoed the river I could identify landmarks he described. Even though Romans did not mention seeing them, the habitat struck me as appropriate for the ivory-bill, and there are historical records of the birds from the area. Here was a southern river that still retained enough of its wildness that I could imagine myself canoeing in the eighteenth century and seeing an ivory-bill fly across the open water to the forest on the opposite shore.

But by the spring of 1972 the river was in peril. There was a great deal of concern in environmental circles over the impending construction of the Tennessee–Tombigbee Waterway, a planned barge canal that was certain to destroy the free-flowing, meandering Tombigbee. The plan would change the meandering southern river—a river so narrow in some places I could throw a stone across and so wide and shallow in others that I could wade across—into a straight canal 300 feet wide and nine feet deep. It would also link the Tombigbee with the Tennessee River, joining the aquatic communities of the two long-separated river systems. I was—and am—convinced that this was wrong. There was tremendous potential for the extinction of some aquatic species that were limited to the Tombigbee drainage. I was also concerned about the clearing of bottomland forests along the Tombigbee. Were there ivory-bills out there? Without the bottomlands I would know the hollow answer. With them, there was hope.

A few weeks later, apparently in response to negative reactions from the environmental community, the Mobile District of the US Army Corps of Engineers held a public hearing on the Mississippi State University campus. The auditorium was packed as speaker after speaker from the Mobile office explained and praised the waterway project. Their mention of wildlife was limited to how the waterway would become a new flyway for ducks and improve the quality of duck hunting in the region.

When the floor was opened for questions at the end of Corps biologist Jack Mallory's presentation, I asked "What about endangered species in the area?" I knew of red-cockaded woodpeckers (*Picoides borealis*) close to the river, had seen bald eagles (*Haliaeetus leucocephalus*) on the river, and one of the last sightings of Bachman's warbler (*Vermivora bachmanii*) was in bottomland hardwoods next to the Tombigbee River in Alabama. Much to my amazement, Mallory said, "We're going to take care of endangered species along the river. Why, I saw an ivory-billed woodpecker just south of Columbus on the river three years ago. We're going to protect those birds so that my grandchildren can see them." I questioned what he said and he repeated that he had seen an ivory-billed woodpecker three years earlier on the Tombigbee.

A few weeks later the environmental impact statement (EIS) for the waterway project was released. When I got it, I turned quickly to the section on endangered species, hoping to find details concerning the ivory-bill sighting. What I found was the following statement: "There are no known endangered species in the project area and there have been no reports of endangered species in the area for the past several decades."

I was outraged!

The Committee to Leave the Environment of America Natural (CLEAN), a local organization, was joined by the Environmental Defense Fund (EDF) in preparing a legal challenge to the EIS. Ultimately the challenge resulted in a court case that was to be heard in district court in Washington, DC. I was asked to testify about the inadequacy of the EIS with regard to endangered species.

When I was called to the stand, I was asked if I had read the EIS section on endangered species. I said that I had, and I quoted what was said. I was then asked if I knew of any endangered species in the area. I did. I spoke of wintering bald eagles, migrant peregrine falcons (*Falco peregrinus*), the red-cockaded woodpeckers, the record of Bachman's warbler, and others. Then the attorney asked if I knew of any other record of endangered species in the area. I said, "Yes, Mr. Jack Mallory, a biologist for the Corps of Engineers, at a public meeting on the Mississippi State University campus, stated that he had seen an ivory-billed woodpecker just south of Columbus three years ago."

Someone in a military uniform in the audience said something out loud in the courtroom—as if in astonishment—and the judge pounded his gavel for order. I was then cross-examined by the government attorney, but quickly dismissed. The ivory-bill record was never mentioned again. I had believed that my mention of Mallory's comments at the public meeting would be dismissed as hearsay, but it was not.

The judge ruled in favor of CLEAN and the EDF and granted a preliminary injunction, stopping construction until the case could go to trial. At that point the lawyers for the pro Tenn–Tom forces argued for a change in venue and won their argument, getting the trial moved from Washington, DC, to the federal court in Aberdeen, Mississippi—where the Tenn–Tom forces had local support. Ultimately the case was lost by the environmentalists. The waterway was built and most, if not all, of the predictions of the environmental groups have come to pass—including that the waterway would not be the economic boon that the Corps predicted and that barge traffic would not come close to the levels predicted.

In an ironic twist of fate, following the 1986 annual meeting of the American Ornithologists' Union, which Mississippi State University hosted, I led participants on a canoe trip down the river. Before the trip I had talked about the former ivory-bill habitat in the area and the recent changes as a result of the Tenn–Tom construction. I had arranged in advance for permission to pass through the Columbus lock on the Tenn–Tom with our canoes. We stopped and talked to the lockmaster before going through. He was ecstatic. Our eleven canoes, he said, were more boats than he had locked through in a long time. We became a statistic for the Corps to use as justification for their destruction. It seemed strange to us that they seemed to make no distinction between a canoe and a barge in tallying traffic on the river.

For my part, I continued to wonder about Mallory's ivory-bill record, and continued searching areas along the Tombigbee River. It was on one of those searches, in March of 1973, when I believe I saw an ivory-bill. I was canoeing down the Noxubee River from Brooksville, Mississippi, to its confluence with the Tombigbee River in Alabama. The Noxubee was at flood stage, making canoeing more difficult because of the strong current and floating logs and debris. At about 10 a.m. and just into Alabama, I saw a large bird fly up from on or near the ground in mature forest. As it flew away from us and parallel to the stream, it appeared to me to be a very large woodpecker, of pileated size or larger, but the trailing edge of each wing was white.

Within a matter of seconds my mind raced to interpret the vision, going from "it was a giant white butterfly" to "no, it's a giant red-headed woodpecker" (*Melanerpes erythrocephalus*), then to "it had to have been an ivory-billed woodpecker." I only saw the bird heading away from me and saw no red. I also heard no calls. Repeated trips to the area over the next several weeks revealed no further sign of the bird. Perhaps I had finally seen one.

On several later occasions as I watched red-headed woodpeckers fly away from me I could see how similar the wing patterns were to those of the ivory-bill. This similarity resulted in a funny incident in the early 1970s. The National Audubon Society sent a writer and photographer to the Tombigbee River to chronicle the beauty and wildness of the region in the days before the waterway was completed. I joined them as they floated down the river and told them of the human and natural history of the region. I mentioned in particular that ivory-billed wood-peckers were known from the area of Tibbee Creek where it flowed into the Tombigbee just south of West Point, Mississippi. They had heard of ivory-bills and were enthralled with the possibility of finding one some-where along the river. How could they be sure of the identity of an ivory-bill if they found one, they asked? I explained that the only sim-ilar bird was the pileated woodpecker and that the ivory-bill could eas-ily be distinguished from it by the large white shield covering its lower back when the wings were folded. I had emphasized how important it would be to photographically document any ivory-bills seen. The next morning I had to return to campus to teach a class. A day later I received a frantic telephone call from my journalist-photographer friends. They said they had found a nesting pair of ivory-bills!

They had spent most of the day photographing the birds using both still and motion picture cameras. They had exhausted their supply of film and wanted me to buy more film and get over there as quickly as possible. It was nearly fifty miles of rural roads, but I was there within an hour and a half—only to find that they had incredible documentation of the nesting of a pair of red-headed woodpeckers. I had assumed that a writer and photographer representing the Audubon Society would be familiar with the relatively common red-headed woodpecker and realize that it was nowhere near the size of the largest woodpecker in America. Alas, their focus was on the key characteristic I had described: the white shield covering the folded wings of the red-headed woodpecker is similar to that of the ivory-billed woodpecker.

During the 1970s I was asked by Lester Short, an ornithologist at the American Museum of Natural History, to serve on the International Council for Bird Preservation's Piciform Specialist Group (woodpeckers being in the taxonomic order *Piciformes*). We were charged with working on problems faced by endangered woodpeckers. My efforts were mostly focused on the red-cockaded woodpecker, but I always found time and energy for the ivory-bill. In late 1985, the US Fish and Wildlife Service appointed me to an Ivory-Billed Woodpecker Advisory Committee, along with Short, Tanner, and several Fish and Wildlife employees. We did not know exactly what the advisory committee was supposed to do, but were told that it was essentially like a recovery team—a term used to define a group of experts that determines how to help an endangered species "recover" to safe numbers. However, we were told we could not call ourselves a recovery team because it was not clear that there was anything to recover.

Our first meeting was held in Baton Rouge at Louisiana State University. When the meeting opened, we discovered why we were there. The Fish and Wildlife Service wanted us to declare that the ivory-billed woodpecker was extinct. Tanner and Short both indicated that they believed the species was extinct in the United States. I was not so sure. Although I had no evidence that it truly existed, I was well aware that there had been no range-wide search for the species since Tanner's dissertation work in the 1930s. The species might be extinct, but there were no scientific data to support such a declaration—and there were literally hundreds of anecdotal reports of ivory-bills. Of course we all knew that most, if not all, of those were misidentifications of pileated, red-headed, or some other woodpecker.

Eventually we all agreed that one last range-wide search was justified. As the one with questions and hope, the search fell to me. I received a grant from the Fish and Wildlife Service for two years of effort—a total of $60,000, almost all of which was for travel. The amount of money did not allow for a field assistant, but I was able to obtain a sabbatical from Mississippi State University, with a reduction in pay of 50 percent. My income would drop, but I knew I was about to begin the greatest field trip of my life.

The committee worked with me to decide what areas were most worth searching and to develop a plan of attack. We also addressed the thorny issue of what should be done if ivory-bills were found. If the location was made public it might spell a quick end to them as a result of an invasion of their habitat by birders wishing to add the species to a life list or by wildlife photographers eager to get a priceless photo. We could not come to a firm decision, but agreed that any such announcement would be made only after the committee had convened to decide on a course of action.

My objectives for fieldwork were outlined during two meetings of the Ivory-Billed Woodpecker Advisory Committee. They were to search for evidence of the continued existence of ivory-billed woodpeckers in the southeastern United States and to evaluate potential remaining habitat for the species.

This book is in part about that search and my searches for the ivory-bill in Cuba, but it is also a chronicle of discovery focusing on this bird, its behavior and ecology, its place in American culture, the reasons for its decline, and the searchers who came before me.

Part One

The Bird

CHAPTER 1

Behavior and Ecology
How It Might Still Live

"Structure is for function."

*W*ithout saying a word, my teacher wrote this phrase on the blackboard. It was the first day of class, and I was a fourteen-year-old freshman at a new high school. I was looking forward to one thing: biology. I loved nature, but I had never really thought about the relationship between an animal's physical structure and its behavior and ecology. "Structure is for function" became a mantra, a mystical formula that opened my eyes to the world around me. It taught me to ask questions. And I applied that questioning to my study of the ivory-billed woodpecker. Why does an ivory-bill have feathers covering its nostrils? How does its barbed tongue work? Why are its tail feathers curved from base to tip? Why are its eggs so shiny and white? My questions seemed endless, but how could I find answers?

Understanding the behavior of a species that is extinct or so rare that it can no longer be studied is a challenge. It requires depending on every shred of evidence available from those who have studied the species, as well as circumstantial evidence from every available source.

I am extremely fortunate to have had James Tanner's published doctoral thesis on the species come along when it did, more than six decades ago in 1942. It may well have been the last hope for really understanding the complexity of the relationship between the ivory-bill and its environment. Tanner's observations play a major role in this first chapter. In a sense these observations are like the cardboard frame associated with a child's jigsaw puzzle: pressed into that frame are the outlines of all the pieces, thus aiding us in getting the picture right.

I felt much like a child as I began studying ivory-bills. To decipher the natural history of the ivory-bill I examined every bit of information I could find, including Tanner's descriptions and photographs, artwork drawn from life, interviews with the few who I truly believe have watched these birds, anecdotal accounts, specimens and pieces of specimens, and inferences from my own knowledge of this and other woodpeckers. With each piece of the puzzle I found, the picture became clearer. To be sure, some of the pieces are missing. And perhaps I have a piece or two upside down or out of place. But I can see a wonderful mosaic of adaptation in this exquisite bird.

A Social Bird?

Bayard Christy suggested in 1943 that the ivory-bill was a social species and that its courtship was communal.[1] I am inclined to agree that they are social, at least as woodpeckers go. In some ways the ivory-bill seems a parallel to the cooperative-breeding red-cockaded woodpecker and perhaps also to the acorn woodpecker (*Melanerpes formicivorus*).[2] For both of these species there is a natural resource that is concentrated and abundant enough that sharing among group members is possible. In the case of the red-cockaded woodpecker, the resource is the cavity. Cavities excavated into living pines take considerable effort to complete and may involve the work of multiple birds over a period of years. Once completed, these cavities are safe nest and roost sites that can be passed on from generation to generation. In the case of the acorn woodpecker, the resource is the store of acorns. The social behavior of ivory-bills might be linked to the large cerambycid beetles these birds seem to favor. Such beetles are found concentrated in large, recently dead trees, and discovery of and access to those beetles may be facilitated by group effort.

The largest reported number of ivory-bills ever seen at one time was eleven.[3] I can hardly imagine the sight of eleven ivory-bills in one scene, but such it once was. Four of the woodpeckers were foraging in the same baldcypress tree (*Taxodium distichum*) during winter, the others nearby, in the late 1800s near Florida's Gulf Coast. Others have often referred to small groups of ivory-bills. John James Audubon, for example, in his journal entry of December 14, 1820, noted that he saw five ivory-bills feeding together. His classic painting of three adult-plumaged ivory-bills feeding amicably together reinforces an emphasis on sociality. In 1935 in Louisiana, George Lowery watched two males and two females feeding on the same dead tree. He wrote, "In manner and disposition the birds are very quarrelsome, although only in a vocal way."[4] Such is also true of the highly social red-cockaded and acorn woodpeckers. In 1948, John Dennis and Davis Crompton observed a group of three ivory-bills, two of which were incubating at a nest, demonstrating this sociability among ivory-bills in Cuba.[5]

Flight

Ivory-bills are known not only to feed together but also to travel together over great distances in their searches for suitable feeding sites. Tanner discussed many aspects of ivory-bill flight, noting that it can vary greatly. His observations were more extensive than those of most other observers and were less anecdotal. From him we learn that ivory-bills have a strong direct flight similar to that of a northern pintail (*Anas acuta*), with steady, rapid wing beats. When traveling any distance, he saw the ivory-bills fly above the trees, thus avoiding navigation among branches. As with other birds, flight ends with a quick, upward swoop, using the wings to brake. Tanner described ivory-bill flight as noisy, the wings producing a "loud, wooden, fluttering" sound as an ivory-bill took flight and beat its wings strongly, and a "swishing whistle" as one flew past.[6] In woodpeckers such as the downy and red-cockaded, wing sound varies with the function of the flight and seems to be an integral part of courtship displays.[7] This could also be the case with ivory-bills.

Tanner's descriptions probably best characterize the ivory-bill's flight, but it seems to me that the diverse descriptions of other authors help demonstrate the variability in its flight behavior. Audubon provided us with considerable commentary on the nature of the flight of ivory-bills. In *The Birds of America* he wrote,

> The flight of this bird is graceful in the extreme, although seldom prolonged to more than a few hundred yards at a time, unless when it has to cross a large river, which it does in deep undulations, opening its wings at first to their full extent, and nearly closing them to renew the propelling impulse. The transit from one tree to another, even should the distance be as much as a hundred yards, is performed by a single sweep, and the bird appears as if merely swinging itself from the top of the one tree to that of the other, forming an elegant curved line.[8]

Audubon often compared the flight of the ivory-bill to that of other species. For example, in his journal entry of November 24, 1820, he noted, "When they Leave a Tree to fly to another they Sail and Look not unlike a Raven."[9]

Others have weighed in with differing views. Lowery described the flight of ivory-bills as like that of other woodpeckers, "undulating and jerky."[10] In contrast, Arthur Allen referred to it as "uniform direct flight" that "resembles that of the red-headed woodpecker more than it does the swooping undulating flight of the pileated."[11]

An interesting comparison can be made between the descriptions of ivory-bill flight and those of the similar-sized great slaty woodpecker (*Mulleripicus pulverulentus*) from southeastern Asia. The great slaty shows similar behavioral and ecological traits that may be a function of the size of the birds. Flight of the great slaty woodpecker has been described as noisy, taking place high above the trees, "often over long distances between successive foraging patches," and as "less dipping than in other woodpeckers and almost crow-like."[12]

Vocalizations and Double Rapping

Other than seeing an ivory-bill flying through the forest, the most common way one might encounter the bird is via its vocalizations or double-rap pounding against resonant wood. My assessment of this piece of the puzzle is based on accounts from observers who truly saw and heard the birds and on recordings made by ornithologists Allen and Peter Paul Kellogg in 1935.[13]

Any attempt to phonetically "spell" or otherwise describe a bird call is fraught with problems: we all hear things a bit differently. Indeed, the same call can also sound differently depending on weather conditions, distance, nature of vegetation between the bird and the hearer, and many other factors. Nonetheless, most of us find mnemonics that work well enough to give a flavor of what we would hear in the wild. Among the mnemonics and other descriptions for the characteristic call notes of the ivory-bill are the following:

Audubon: "*pait, pait, pait*" and also like a child's "tin trumpet."[14]

Frank M. Chapman: "a high, rather nasal *yap, yap-yap.*"[15]

Allen and Kellogg: "*kent.*"[16]

R. D. Hoyt: a weak "*schwenk.*"[17]

S. C. Graham: "the burry reed notes of the bagpipe."[18]

Lowery: "loud, clear, high-pitched '*yaamp-yaamp*'" with a "decided nasal twang" and "rather plaintive."[19]

T. Gilbert Pearson: not as loud as those produced by several smaller woodpecker species; nasal *kent* calls almost continually uttered while feeding."[20]

Audubon's first language was French, causing me to wonder if variance between Audubon's mnemonic devices and those of other authors may be a result of his phonetically spelling the mnemonic descriptions as he would in French. Audubon rarely provided mnemonics in his species accounts, but when he did they were unusual relative to mnemonics provided by others. As another example, he described the familiar *"Who cooks for you, who cooks for you'all"* of the barred owl (*Strix varia*) as *"whah, whah, whah, whah-aa."* In any case, we are incredibly fortunate to have the recordings of Allen and Kellogg. Without them, our foundation for discussing ivory-bill vocalizations would be shaky indeed. Recordings can distort calls and songs. Yet careful attention to these recordings suggests that they are accurate because the sounds of other birds in the background, particularly common and well-known birds, do not sound distorted. A red-bellied woodpecker (*Melanerpes carolinus*) in the background on one recording sounds as clear as one that we might hear in the forest today.

The recordings support the descriptions of Audubon and others who likened the ivory-bill's call notes to the sound of a child's tin trumpet. Allen noted the call of the ivory-bill, as recorded, could be easily imitated by using a clarinet mouthpiece. Shortly after my arrival at Mississippi State University in 1970, James Tate, then assistant director at the Cornell Laboratory of Ornithology, sent me a gift to celebrate my new position: an old clarinet mouthpiece that he had found at the lab, perhaps one used by Allen or Tanner. Tate recognized that I was in "ivory-bill country" and encouraged my searches. I found that by blowing forcefully on the mouthpiece, I could imitate closely the calls of the ivory-bill on the Allen recordings.

Audubon provided us with considerable description of ivory-bill calls, but his most important information may have to do with context and diversity of calls rather than on the mnemonic descriptions. Audubon commented that the ivory-bill "never utters any sound whilst on the wing, unless during the love season; but at all other times, no sooner has this bird alighted than its remarkable voice is heard at almost every leap which it makes while climbing."[21]

Audubon's comment about not calling while in flight has been disputed. E. A. McIlhenny, who was both a southern naturalist and founder of the Tabasco Sauce industry, lived at Avery Island in south Louisiana, not far from Audubon's former home.[22] McIlhenny found ivory-bills to be exceedingly silent and noted that he never heard them *except* when they were flying. George Beyer, another Louisianan, saw and heard ivory-bills along the Tensas River, and agreed with Audubon's characterization, adding that "if unmolested and not alarmed, they are certainly noisy."[23]

Tanner elaborated on some of the variation in the ivory-bill's vocalizations, noting that when disturbed the pitch of the *kent* rises and is repeated more rapidly, often as a double note.[24] This is readily heard on the Allen-Kellogg tapes. When two or more birds were together, particularly just before taking flight, Tanner noted that they gave a "chorus" of prolonged, upward-slurred *kient, kient, kient* notes. When mated birds were together during the breeding season, they often exchanged lower, softer notes that Tanner described as *yent, yent, yent*. Allen and Kellogg also described the note of the female as being weaker and less harsh, but Tanner could not discern this difference.[25] Audubon referred to what seems like social chatter among birds foraging close to one another in his journal entry of December 14, 1820: five ivory-bills feeding on the berries of "creepers" were "gentle" and kept up a "Constant Cry of *Pet Pet Pet*."[26] Dennis also described the birds as having a "conversational chatter."[27]

Tanner maintained that the ability to hear ivory-bills at a distance depended on the notes given, where they were given, and the season of the year. Under the best of conditions—on a calm day in winter when the leaves were off the trees—he could hear an adult giving the loud *kient, kient, kient* call from more than a quarter mile away.

Is one person right or another wrong with regard to the nature, frequency, volume, and diversity of ivory-bill vocalizations? I suspect not. Like other woodpeckers, there was probably considerable seasonal variation in the frequency and nature of calling. The frequency of calling was also likely influenced by social circumstances. In the red-cockaded and red-headed woodpeckers, for example, solitary birds are often much less vocal than birds in a family group. I suspect that these observers were each focusing on impressions from their own experience and, in the case of McIlhenny and Beyer, that was somewhat limited. From my perspective, the extensive experiences of Audubon and Tanner should be given greater weight.

Calling is one type of sound produced by woodpeckers, but the sounds most people associate with this group of birds are the raps of a bill against a tree or other resonant surface. Most woodpeckers drum rapidly and with a unique cadence. There have been a few accounts of "drumming" by ivory-bills from the nineteenth century.[28] McIlhenny recorded that "one bird will alight on a dry limb of some tree and rap on it with its bill so fast and loud that it sounds like the roll of a snare drum; this it continues to do at short intervals until its mate comes."[29]

Although drumming is a characteristic of many woodpeckers, it is generally thought to be absent in the *Campephilus* woodpeckers. Thus, if ivory-bills do drum, they are probably the only members of their genus that display the behavior. It seems more likely to me that they do not, and instead use a double rap as their method of communicating. This double rap seems to be characteristic of woodpeckers in the genus *Campephilus*.[30] Tanner described this mechanical sound as a heavy pounding with the bill on a limb or stub: sometimes a single blow is given, but more often the bird gives "a hard *BAM-bam*, the second note sounding like an immediate echo of the first."[31] He demonstrated the double rap for me at an advisory committee meeting by rapping his knuckles on a table, and the echo impression was clear.

Tanner, Audubon, and others have suggested that ivory-bills sometimes give their loud double rap when disturbed or in response to calls or raps from other ivory-bills, or even in response to the blows of a woodsman's ax. When I am in the field I often use my own tapping to attract woodpeckers—not drumming but the kind of tapping often associated with foraging. I have found that different species readily respond to such efforts and that the cadence needed to elicit a response can vary from species to species. A pocketknife tapping against a large plastic tag attached to a motel key cupped in my hand has been especially effective for red-bellied woodpeckers. A key tapped against a quarter cupped in the palm of my hand has worked for downys. The signal value of woodpecker tapping is clear. The double raps of an ivory-bill may serve as a warning signal or perhaps at times as a location signal, letting others in the group know of its whereabouts.

Feeding Adaptations

As you might expect from a bird capable of producing such a loud, forceful rapping, the ivory-billed woodpecker's bill is incredibly robust, achieving remarkable strength through a combination of its straightness and broad base. Measuring about three inches long and tapering to a flat chisel tip on both the upper and lower bill, it can serve not only as a percussion instrument but also as a wood chisel, probe, dagger, and pincers. The slitlike nostrils of the ivory-bill are located near the base of the upper bill and are hidden beneath whitish bristlelike feathers that stretch forward. These feathers may act as a filter to prevent the wood chips and dust that fill the air as the woodpecker excavates from entering their lungs.

The long toes of the ivory-bill give it a broad, solid grip on tree surfaces, one that allows it to stay perched on the side of the tree even while large chunks of wood are chiseled away. The third toe is the longest and normally reaches out to the side, providing stability as well as grip. Each toe is equipped with a large, curved, sharp claw—almost a semicircle. The tip of each claw contacts the surface perpendicularly, much like the curved ice tongs used to grip a large block of ice, thus effecting the best possible grasp of a difficult surface.

The way in which an ivory-bill moves on tree surfaces is important to the manner and efficiency with which it obtains food. Although some have noted that it hitches up tree surfaces in a manner similar to that of other woodpeckers, the movement, as described by Tanner, is a springing upward, the legs simultaneously pushing off and stretched back, then brought forward again toward the end of each leap—a "quick and vigorous, almost nervous" action.[32] Again, structure and function come together in a unique way with the ivory-bill, because its tail is highly modified to assist in this movement. Examination of the tail feathers of an ivory-bill reveals not only the characteristic woodpecker strength and stiffness but also curvatures unlike the tail feathers of other woodpeckers. From the base to the tip, each feather is curved downward (Plate 5). And from the midrib, the vane on each feather is also curved down such that in cross-section toward the tip of a tail feather, the feather is nearly U-shaped. These curvatures together with the feather strength and stiffness create a type of feather "spring." As the ivory-bill hits a tree surface with its bill, the recoil pushes against the tail feathers, "cocking the spring," which then pushes the bird forward, contributing to the force of the next blow. The upward movement of an ivory-bill would be similarly benefited by this spring, hence the springing motion noted by Tanner.

Foraging Behavior

Most observations of foraging ivory-bills have been made at recently dead trees. In general, feeding birds knock off large slabs of bark to obtain larvae beneath. In a letter to Tanner, Herbert Stoddard, who knew ivory-bills in his youth, described what he believed was ivory-bill work on pines that had been killed by a hurricane in north Florida: "The larger portion of the bark of these pines had been removed while it was still quite tightly attached, the evidence being left on the tree being comparable to that a man might leave who knocked off the bark with a cross hatching motion with a heavy screwdriver."[33]

This description would easily fit the work Tanner described on hardwood trees in the Singer Tract and some that I saw on pines in the mountains of eastern Cuba. Allen thought that there might be geographic or habitat differences in the foraging ecology of ivory-bills, noting that although the birds nested in a baldcypress swamp, "They did most of their feeding along its borders on recently killed young pines that were infested with beetle larvae."[34] Ivory-bills have also been noted to occasionally come to the ground in recently burned areas, and once there move around by "hopping like a flicker."[35] Rather than illustrating geographical differences in their foraging, I believe Allen's observations illustrate an opportunistic flexibility in the ivory-bill's foraging.

Ivory-bills characteristically forage in pairs or as family groups, often on the same tree or even the same limb. Other species, including the pileated woodpecker, may forage commensally with ivory-bills (as hairy woodpeckers do with pileated woodpeckers), eating arthropods that are uncovered by the ivory-bill's work.[36]

Ivory-bills probably wander over large areas in search of recently dead trees that might harbor their favored insects. Audubon noted that when settlers girdled trees to kill them as they were trying to clear areas to farm, the ivory-bills were attracted to them.[37] This certainly supports the idea that one might girdle a few trees in areas where ivory-bills are suspected to occur and thus lure them in for confirmation of their presence. It also suggests a method of managing habitats, at least over the short-term. One of the first tasks, if we locate ivory-bills, will be to increase numbers. Thus girdling trees might aid in providing the food resources needed to feed an expanding population.

The Ivory-Bill's Diet

I have talked about the importance of beetle larvae in recently dead trees to the diet of ivory-bills, but it would be misleading to suggest that that is all they eat. Most of what we know about ivory-bill diet is based on observations of their foraging behavior, but some records of stomach contents exist. In 1911, F. E. L. Beal examined the contents of two ivory-billed woodpecker stomachs and found that one contained thirty-two and the other twenty cerambycid beetle larvae.[38] The larvae constituted 37 percent of the food in the stomachs. Also in one of the stomachs were larvae from engraver beetles, three species from the family Scolytidae. Animal food made up 38.5 percent of the contents of the two stomachs and vegetable matter the other 61.5 percent. One stomach contained the fruit of the southern magnolia (*Magnolia grandiflora*) and the other stomach included pecan (*Carya illinoensis*) nuts.

In 1939, Clarence Cottam and Phoebe Knappen reported on the examination of the contents of three ivory-billed woodpecker stomachs, apparently the first two of which were the same ones examined by

Beal.[39] Cottam and Knappen reported that the first two stomachs came from birds collected by Vernon Bailey on November 26, 1904, near Tarkington, Texas. The third stomach came from a bird collected at Bowling Green, West Carroll Parish, Louisiana, on August 19, 1903, by E. L. Mosely. They noted that the first two stomachs were filled and that they did not actually have the stomach of the third bird but had received only its contents. The combined sample from the three stomachs included 46 percent made up from animals, and most of that was from cerambycid beetles. Two species of cerambycids were identified from the three stomachs. The first was *Parandra polita*, a long-horned beetle that has been described as rare in the southern United States but common in Mexico and Central America, thus potentially providing a specific dietary link between the ivory-bill and Mexico's imperial woodpecker (*Campephilus imperialis*).[40] These beetles feed on the heartwood of old and weakened hardwoods trees. The other beetle, the hardwood-stump borer (*Mallodon dasystomus*) is found only in the eastern United States, and its larvae consume the heartwood of living trees.[41]

Cottam and Knappen broke down the plant component of the diet as including 14 percent southern magnolia seeds, 27 percent hickory (*Carya* sp.) and pecan nuts, and 12.67 percent poison ivy (*Toxicodendron radicans*) seeds. They also noted that a trace of gravel was in the third stomach and that the stomach contents also included fragments of an unidentified gall. Audubon and others also described considerable use of fruit and berries in season, noting that persimmons (*Diospyros virginiana*) and hackberries (*Celtis* sp.) were eaten and that ripe grapes (*Vitis* sp.) were especially favored.

Although the favored food of ivory-bills appears to be the large larvae of some long-horned beetles of the family Cerambycidae, the birds have also been reported to be attracted to trees killed during what appear to be outbreaks of southern pine beetles (*Dendroctonus frontalis*). Southern pine beetles are members of the family Scolytidae and have larvae that are rice-grained-size, much smaller than the fat, 2 1/2-inch larvae of the cerambycid beetles. Although it seems probable that ivory-bills would feed on southern pine beetle larvae, I cannot help but wonder if they would not strongly prefer the larger beetles that provide much bigger "units" of energy for the effort. This is one of those missing puzzle pieces: there is just not enough information to resolve this question with certainty. Although scientists' emphasis on cerambycid beetle larvae as a major food of the ivory-bill is justified by the data available, there *are* few data. Reports of ivory-bills foraging on downed wood suggests the probability that they also readily took the large larvae of other, as yet undocumented, beetles, such as the horned passalus (*Popilius disjunctus*).

The ivory-bill's apparent preference for the large larvae of cerambycid beetles seems likely to have led to the scientific name of this woodpecker. The genus name *Campephilus* means "caterpillar lover." Although beetle larvae are certainly not caterpillars, these large white beetle larvae have a size and form much like the caterpillars of some larger moths. I can easily imagine that on opening an ivory-bill's stomach and discovering such larvae that they might be believed to be large caterpillars bleached by stomach acids (see Plate 3).

The diet of the ivory-billed woodpecker in Cuba is probably similar to that of the birds in the southeastern United States. George Lamb identified both small and large beetle larvae from limbs on which he had just observed an ivory-bill foraging, and while in Cuba, I collected large cerambycid beetle larvae from pines on which ivory-bills had been foraging.[42] Although the ivory-bill does not seem to be an ant or termite eater, as many woodpeckers are, there are reports from Cuba that they do feed on arboreal termites and seeds.[43] It seems certain that the diet of the ivory-billed woodpecker varies regionally, among habitats, and seasonally. As with other species, the ivory-bill seems to be an opportunist, taking advantage of varying food resources and responding to changing needs to satisfy energy demands for molting, courtship, and reproduction.

Courtship and Nesting

All we know of the courtship and nesting ecology of ivory-billed woodpeckers is anecdotal—bits and pieces scraped together from the comments of a handful of observers. The best information we have comes from the observations of Allen, Kellogg, and Tanner. We can continue to learn new things from their observations, especially from the photographs provided by Allen and Tanner, but I also feel the need to speculate on what is not known. By using our understanding of the behavioral ecology of other woodpecker species, particularly other species in the genus *Campephilus*, we can try to fill in more missing puzzle pieces.

Virtually nothing is known of the courtship of the ivory-bill. During exchanges at nests, both Allen and Kellogg and Tanner observed clasping of bills.[44] Allen's photo illustrating an exchange at the nest (Figure 1.1) nicely shows the posture of the birds as they approach one another. It also reveals something that neither Allen nor Tanner mentioned: the bird leaving the cavity stretches up in an awkward position as it exits and the secondary feathers are bent against the tree surface. I had not noticed this in the photo until I examined many specimens that show exceptional wear on those same feathers. When I checked the dates of the collection of the specimens, I realized that the birds were likely nesting. Then I happened to notice Allen's photograph and quickly realized how those particular feathers had likely become worn—in a nest exchange that takes place at the nest cavity.

There is little information that helps us see any pattern to the timing of ivory-bill nesting. This is a species that probably nested in at least a dozen states, from North Carolina, Kentucky, and Missouri, south to the Gulf Coast and west to Texas and Oklahoma. Yet for all the specimens we have—many of which were shot at the nest—we have nesting records from only four states. All ivory-bill specimens were of birds shot intentionally—some by scientists, most by hobbyists or simply curious hunters. Birds continued to be shot for collections long after it was realized that they were in danger of extinction.

Tanner lists fourteen nest records from Florida, three from Georgia, sixteen from Louisiana, and two from east Texas. The earliest nest record is for an adult reported incubating on January, 20, 1880, by W. E. D. Scott.[45] Scott mentioned taking birds at two localities: Panasofkee Lake

Figure 1.1. The behavior of a pair of ivory-bills as the female arrives at the nest to take the male's place during incubation in Louisiana's Singer Tract, April 12, 1935. The birds waved their raised bills back and forth in a mock "duel" during the exchange. Note how the white inner secondaries of the male's wing are pushed against the tree surface during the exchange, potentially accounting for the extreme wear seen on these feathers on specimens that were collected during the nesting season. *Courtesy Arthur A. Allen II*

and at the mouth of the Withlacoochee River, and it is not clear which of these was the locality for the incubating female, although both localities are near the Gulf Coast, west of what is now Leesburg and north of Tampa. Although Tanner accepted Scott's record, it was exceptionally early in the year, and Scott's mention that this bird was incubating might have meant only that she was shot at a cavity. Scott made no mention of seeing or collecting eggs, nor does he mention collecting a male at the nest. It seems likely that he would have collected the male if it had truly been a nest, because the nest cavity is the male's roost cavity and he would share incubation during the day. On the other hand, Hoyt found two eggs in a nest near Clermont, Lake County, Florida, on February 15, 1905, and noted that incubation "was well under way."[46] On March 9 of the same year there was a second nest about seventy-five yards from the first, again with two eggs.

Hoyt reported that, in Florida, nest-cavity excavation takes place in late January, with eggs normally laid by February 10, and young leave the nest in April.[47] This scenario fits fairly well with our understanding of other North American woodpeckers. In general, cavity excavation takes about two weeks, laying takes two to three days (one egg per day), incubation lasts about two weeks, and young fledge in about twenty-eight to thirty days.[48] Because ivory-bills are larger, their nestling life might be slightly longer than that of smaller species, but I know of no evidence that suggests that the length of a successful ivory-billed woodpecker nesting effort, from cavity excavation to fledging, differs greatly from that of other woodpecker species.

Initiation of nesting begins slightly earlier in the south in many species. We have little evidence to demonstrate such a pattern in the ivory-bill because of a lack of information for nests from more northern parts of its range. However, southern populations began nesting when it would have been quite cold at more northern latitudes. In south Florida, Scott mentioned that ivory-bills nested in early February.[49] In west-central Florida they had eggs by about mid-February. In northeast Louisiana, Tanner reported small young as early as February 17.[50] Tanner also found small young as late as May 10 to 13. The latest egg date known is for a clutch of three eggs found on May 19, 1892 (a second clutch) on Avery Island, Louisiana, by McIlhenny.[51] Beyer found a well-feathered nestling in Louisiana in July 1899.[52] Tanner also noted a range of at least two months between fledging of the earliest young and fledging of the latest in a population. This range of times likely includes renesting efforts. Renesting seems to occur if the first clutch fails, because there is no evidence that ivory-bills ever produced two broods in a season. Beyer observed ivory-bills in July in Louisiana and commented that he did not believe that they had a second brood, because even in July the adults were much too busy providing attention to the fledglings.

Reported clutch sizes range from one to six eggs and brood sizes from one to four young. Harold Bailey reported a set of four eggs taken by S. W. Wilson in Georgia's Altamaha Swamp on April 10, sometime before 1865.[53] Maurice Thompson reported a clutch of five eggs for a Florida nest.[54] Audubon reported clutches of six eggs, but provided no details.[55] Tanner presented evidence that earlier nests had fewer eggs, but sample sizes were small, making conclusions difficult.[56]

We have no data on either the onset of incubation or the incubation period of ivory-bills. Furthermore, I have been able to find no information on the timing of incubation in any other *Campephilus* woodpecker—clearly a gap in understanding of these large woodpeckers. From other woodpeckers we know that incubation usually begins with laying of the last egg. However, because the nest is in the male's roost cavity and he is there each night during egg-laying, his body heat seems to be enough to sometimes initiate development before laying of the last egg. Whether true incubation (the deliberate application of heat to the eggs) begins before laying of the last egg is not known, but at least in some nests development can begin earlier in earlier eggs and hatching can thus be staggered over about two days. This happens fairly frequently, and the earliest hatched young can double their weight during their first day, thus gaining a competitive advantage over late-hatched young. In such cases, during times or at sites when food is not readily available, brood reduction can occur through loss of the smallest chick.

Tanner suggested that the incubation period of ivory-bills might be about twenty days.[57] Woodpeckers in general have short incubation periods. Red-cockaded woodpeckers, for example have an incubation period of about ten to eleven days.[58] Incubation period tends to be longer in larger species, but twenty days for ivory-bills seems too long. It is true that an incubation period of eighteen days has been given for the pileated woodpecker, but on minimal evidence.[59] Lawrence Kilham, who bred pileated woodpeckers in captivity, found their incubation period to be only fifteen to sixteen days.[60] The incubation period for the similar-sized black woodpecker (*Dryocopus martius*) is considered to be twelve to fourteen days.[61] Without additional data, we will never know the true story of the ivory-bill's incubation period.

Early mention of nestlings was made by collectors who were looking for eggs or who were interested in skins of adult birds. If they climbed to a nest in search of eggs and found nestlings, they were disappointed, usually did not bother to comment on their find, and in some cases may have destroyed the young in hopes that it would induce the adults to re-nest so that they could obtain fresh eggs. However, we do have a few descriptions of young. The nest Scott found near Tarpon Springs, Florida, included a chick he described as about one third grown, slightly feathered, eyes not yet open, feathers of first plumage "beginning to cover the down," and "exactly the same in coloration" as the adult female.[62] This is information that tells us several things. First, the occurrence of yet closed eyes as the feathers of the first plumage are emerging is as we see in other woodpecker species. In all woodpeckers, young are hatched naked and with eyes closed and covered by translucent eyelids. Nestlings can sense changes in light and beg in response to the darkening of the cavity as a parent arrives at the entrance.[63] By about the seventh day after hatching, the feather tracts are dark and just the tips of feathers may be poking through. By the tenth day the tip of many feathers has emerged and opened, but the eyes are usually just beginning to open. Based on this knowledge, my estimate of the age of Scott's nestling would be about nine to ten days. Being a larger bird, development might have taken longer, so that perhaps the chick was as old as eleven or twelve days. Contrary to Scott's description, no woodpeckers are known to have natal down, and it is highly unlikely that ivory-bills did. The first plumage of woodpeckers (and other birds) is characteristically weaker, lacks any iridescence that adults might have, and is fluffier in appearance as a result of the presence of fewer barbs on the vanes. More durable plumage is not needed because this first plumage is replaced quickly; often replacement begins even before the young leave the nest.

Allen and Kellogg described the begging calls (as heard from the ground) of small nestling ivory-bills as a "weak buzzing."[64] Tanner described them as "a rapidly repeated *chirp-chirp-chirp*" moderately pitched and "rising in pitch and intensity."[65] As nestlings matured, they sounded increasingly like adults. Both parents brought food to nestlings, and from one observation period to another the sex bringing the most food differed. Overall, however, the contributions of the sexes were close: Tanner observed seventy-seven feedings by males and eighty-two by females. He also noticed a feeding pattern that I have observed in other woodpeckers: a rush of feedings early in the morning, a lull at midday, and a flurry again in late afternoon. Such a pattern is easily understood. Nestlings must fast all night and by morning are hungry, hence the flurry of early-morning feedings. Once the nestlings are fed, the adults must feed themselves, hence the decline at midday. It is then important to stoke the nestlings just before they begin their nightly fast.

Beyer examined a nest cavity shortly after a single ivory-bill nestling had fledged.[66] He found the interior of the cavity "very clean," noting "every sign of excreta and other household debris had been carefully removed." Tanner found that every nest he examined was kept clean but that a few dried white droppings were mixed with chips at the bottom of nests from which young had fledged. He observed only the male removing fecal material from the nest. The material was sometimes dropped at the nest entrance and sometimes carried a short distance and dropped in flight. It makes sense that the male would keep the nest clean, because it is his roost each night.

Nest Sites and Cavities
Tanner summarized reports of at least thirty-five specific nests and

wrote that at least twelve species of trees were used as nest sites.[67] Eight of the nests were reported in baldcypress, two in pines (*Pinus* spp.), two in red maple (*Acer rubrum*), and four in oak species [including one each in Nuttall oak (*Quercus nuttalli*) and overcup oak (*Quercus lyrata*)]. Other trees reported as used for nests included one nest each in "bay" (Magnolia?), white elm (*Ulmus americana*), sweetgum (*Liquidambar styraciflua*), tupelo gum (*Nyssa aquatica*), "ash" (*Fraxinus* sp.), "hackberry," and cabbage palm (*Sabal palmetto*). Audubon suggested that nests were generally in ash or "hagberry" (hackberry); Scott suggested that many nests were in baldcypress, but also that palmetto [likely cabbage palms] trunks were also commonly used.[68] Unfortunately, in many cases we do not know if the tree was live or dead, how big it was, or where the nest was relative to the lowest branches. In cabbage palms, nests were apparently as low as fifteen feet, clearly reflecting the palm's smaller stature.[69]

Audubon suggested that ivory-bills never nested in dead trees, but there are now many records to the contrary, including photographs of their use of dead snags as nest trees. At least six of the thirty-five nests reported in Tanner were in dead trees.[70] In Louisiana, McIlhenny thought ivory-bills preferred a "partly dead" tree, but he noted that they always nested in the dead part—never where there was gum flow, but also never where there was extensive rot.[71] Excavation of a cavity in sound wood is a difficult and time-consuming task that woodpeckers generally avoid. Instead they take advantage of wood softened by wood-decaying fungi, but probably not extensively rotted wood. Wood that is excessively rotted would provide easy access to many predators: it affords a rough surface that allows climbing predators to grip, and it would be easy to rip open.

In a live or dead tree, the nest site was either in a dead stub or below a dead stub. The focus on the word "stub" is important. A broken off tree or branch exposes a cross-section of the grain of the wood to rain and heavy dew. This moisture seeps into the wood, following the columns of xylem cells, more rapidly penetrating the tissues than water seeping in across the grain. Such moist wood provides a perfect environment for wood-rotting fungi whose spores and tissue fragments are carried on the wind (and no doubt on the bills of woodpeckers). Once the fungi take hold, insects such as termites and some beetles invade, consuming both the fungi and the rotted wood. Woodpeckers take advantage of the fungal decay to more easily excavate their nest and roost cavities.

A branch or stub above a woodpecker nest does several important things: it shields the cavity entrance to some extent from both sun and rain; it will often cast a shadow on the cavity entrance, making the entrance more difficult to detect; and if the branch is a dead stub, it provides an avenue for fungi to enter the tree and soften the heartwood, thus making it easier for the woodpecker to excavate a cavity at such sites. Audubon wrote of the tendency to place the nest under a stub as follows:

> The birds pay great regard to the particular situation of the tree, and the inclination of its trunk; first because they prefer retirement, and again, because they are anxious to secure the aperture against the access of water during beating rains. To prevent such a calamity, the hole is generally dug immediately under the junction of a large branch with the trunk.[72]

A cavity provides a nest or roost that is protected from the weather and from many predators—but only as long as it is sound. Most woodpeckers make use of a cavity for a single season or year. Harold Bailey suggested this is true of ivory-bills as well, although, like other woodpeckers, nests in subsequent years are often nearby, even in the same nest tree.[73] As nestlings mature, the efficiency of nest sanitation by the parents of most woodpeckers becomes less and less. Excrement, fragments of eggshells, food, and feathers accumulate on the bottom. Some rain often enters the cavity. All of this provides a fertile environment for the growth of fungi, bacteria, and a host of small arthropods and other creatures. The decay process continues and is enhanced by the woodpecker's activity. All of this also provides a habitat for the diversity of lice, mites, and other external and internal parasites that the bird may harbor. At first the brooding adults or roosting male may chip at the walls of the cavity, adding clean chips to the bottom, but this activity soon ceases. The nest cavity of most woodpeckers is a mess at fledging, and increased fungal growth makes cleaning and rehabilitation of the cavity uneconomical, although ivory-billed woodpeckers may have been somewhat more fastidious about nest sanitation.

A few woodpeckers do reuse nest cavities. Red-cockaded woodpeckers, for example, characteristically reuse their nests. The red-cockaded nests in the trunk of living pines and thus its nest cavity is usually surrounded by sound wood. Such cavities do not deteriorate as rapidly as those in dead stubs. Others, such as the hairy woodpecker, that occasionally use live limbs or trunks for cavity excavation will sometimes reuse a cavity.

Scott suggested that the ivory-bill nest he found near Tarpon Springs, Florida, had been used as a nest the previous year.[74] He was guessing on the basis of the fourteen-inch depth of the cavity. He might have been right, but his claim lacks documentation. Whether woodpeckers reuse a cavity or not, they often reuse the same nest tree, excavating a new cavity below one previously used. Continued downward progression of decay often results in a succession of cavities being excavated, usually below that of the previous year.

No ivory-bill cavity entrance orientations have been recorded, except that photos and descriptions indicate that most cavities open in the direction the nest limb leans. This characteristic, common to woodpeckers and other cavity-excavating birds, reduces the potential for the cavity to fill with rainwater and may also shelter the cavity to some extent from aerial predators. The nest cavity of an ivory-bill might also need to face in a direction that was relatively open to facilitate the arrival and departure of such a large bird (Figure 1.2). Recorded heights of ivory-billed woodpecker nests have ranged from about fifteen feet to more than seventy feet, averaging higher than thirty feet.

In some woodpecker species, such as the red-cockaded woodpecker, authors have suggested that the birds avoid sites where vegetation comes close to the cavity entrance, arguing that such vegetation might facilitate

Figure 1.2. Photo of an ivory-bill nest in the Singer Tract in northeast Louisiana. Note the relative openness of the surrounding forest and the branch stub just above the cavity entrance. Such branch stubs not only offer some shelter from above but also provide an avenue for the entrance of fungi that soften the heartwood and facilitate cavity excavation. *Courtesy James T. Tanner*

access to the cavity for a predator. Some descriptions of nests demonstrate that this was not the case with ivory-bills. Beyer found a nest entrance nearly obscured by a poison ivy vine, and Tanner, after replacing a nestling in its nest, pruned a branch that was obscuring his view of the nest.[75]

Audubon first described the oval nature of the entrance to ivory-billed woodpecker cavities as being taller than they are wide. Beyer seems to contradict this description, noting, "The entrance to the nests never seems to be circular as that of other woodpeckers, but is a little wider than high."[76] But then he provided an exact measurement of the cavity entrance as 4.25 inches high and 3.88 inches wide—in apparent agreement with Audubon. I suspect that the greater height of the ivory-bill's (and often also pileated's) oval cavity entrance functions to accommodate the crest of an entering bird.

In addition to the vertical and horizontal dimensions of the entrance, Beyer also suggested that the oval shape of the ivory-bill cavity entrance "is greatly augmented by the peculiar and ingenious way of slanting the lower edge of the hole for the purpose of shedding the rain."[77] Such beveling is a common characteristic of red-cockaded woodpecker cavities, and I have seen it to a lesser extent on most woodpecker cavities. The effect is that the lower lip of the cavity is recessed somewhat in comparison to the upper lip and that the vertical face of the cavity below the entrance slopes outward to meet the contour of the tree surface. Water coming down from above thus hits the outward-sloping surface and is carried away from the entrance rather than accumulating or even flowing inward. I suspect, however, that this is not an ingenious method of shedding the rain but primarily wear of the decayed wood from the repeated entrance and exit of the birds—it is a well-worn path to their home, and this path incidentally has this beneficial effect.

Nests have been described as 14 to 25 inches deep and up to 10.5 inches wide.[78] Beyer reported a nest "nine inches deep and seven and a half inches in height," but this is difficult to interpret.[79] No author clearly describes the measurements taken, and thus we really do not know cavity dimensions. Beyer might have measured the cavity as 9 inches from the bottom lip of the entrance to the bottom of the cavity and 7.5 inches from the bottom lip of the cavity to the top of the cavity.

Eggs and Nestlings

As with all known woodpeckers, the nest cavity of ivory-bills is neither lined nor filled with nesting material. Instead, a few chips are left on the bottom of the cavity, and eggs are laid on these (Plate 1). During incubation and early development in other woodpeckers, one or both of the parents may chip away at the inside of the cavity, providing fresh chips. This may be a result of fidgety boredom on the part of the incubating or brooding adult, efforts to increase the dimensions of the cramped cavity for the comfort of the adult, or after hatching, to provide a clean substrate to cover excrement. Adding more chips to the bottom clearly accomplishes the latter in red-headed woodpecker nests I have observed.

Fledglings and Fledgling Dependency

Leaving the nest is a major event in nestling life and one for which the triggering stimulus is often not known. For ivory-bills, the frequency of feedings decreases as the time of fledging approaches. Tanner observed a period of three hours and forty minutes without a feeding at midday on the day before one nestling fledged. This is characteristic of woodpeckers, and these reduced feedings—accompanied by increased activity of nestlings at the nest entrance and stretching out to retrieve food from an arriving parent—seem to be a prelude to fledging. As demonstrated by Tanner when he tried to return the banded chick to its nest, premature fledging can be caused by disturbance at the nest.[80] Broods of four young ivory-bills are known to have fledged, although most fledged broods include only one or two young.

Anecdotal descriptions of foraging groups of ivory-billed woodpeckers by early naturalists in the United States and Cuba suggest an unusual social system or an extended period of juvenile dependence. Tanner documented young remaining with their parents until the beginning of the next breeding season. At that time, however, he observed at least one instance where the breeding female repeatedly tried to drive a young male away, while the breeding male seemed to ignore him. Audubon noted differences between younger and older birds that could be distinguished beyond their second year, noting that they did not attain "their full size until the second year. Indeed, even then, a difference is easily observed between them and individuals which are much older."[81]

A study skin of an ivory-billed woodpecker in adult plumage but with reduced bill development suggests delayed bill maturation.[82] A

Figure 1.3. Side view: The juvenile female's *(below)* bill is more pointed in side view, less chisel like, and less suitable for scaling bark from trees in the manner characteristic of adults. Shaping of the bill as an efficient tool for excavating insect larvae is likely a function of both growth of the bill and wear. The larger bill of an ivory-bill probably necessitates a much longer period of maturation than the smaller bill of other woodpeckers. *Courtesy Jerome A. Jackson*

juvenile ivory-bill appears to have a bill that is both shorter and more rounded at the tip (Figure 1.3, Plate 6). A young bird without a fully developed bill may have limited foraging abilities, resulting in a longer period of dependence on its parents.

As with other woodpeckers, the plumage of juvenile ivory-bills is more loosely webbed (fewer barbs per inch on the rachis—i.e., shaft of the feather) and duller than that of adults. It shows none of the slight iridescence found in adults. The four large outer primaries (the sixth through ninth) are rather attenuated in juveniles and less flattened at the tip than in older birds. The juvenal tenth primary of fledglings is much longer and more rounded than that of older birds and can probably be used to identify juveniles through their first summer. The postjuvenal molt seems to occur between June and October, and the white secondary feathers seem to be retained.

One of the peculiarities of ivory-bills is that the juvenal plumage of males lacks red in the crest. Audubon first described the juvenal plumage of ivory-bills and the transition to adult plumage, noting, "The young are at first of the color of the female, only that they want the crest, which, however, grows rapidly, and toward autumn, particularly in birds of the first brood, is nearly equal to that of the mother. The males have a slight line of red on the head and do not attain the richness of plumage until spring."[83]

Tanner noted that the first red feathers appeared in the crest of his banded juvenile at about two and a half months after fledging and that at that time the crest was still blunt and ragged in appearance. This banded male in juvenal plumage had white tips to some of the outer primaries, and Tanner felt that these might also characterize young birds (Figure 1.4).[84]

The bill of a juvenile is not only shorter and more rounded at the tip (see the earlier discussion of fledgling dependence), but it is also paler.[85] Tanner described the eye color of juveniles as dark sepia. The eyes of his banded male were taking on a lighter appearance by the age of two and a half months.

Ivory-Bill Predators and Competitors

As large as ivory-bills are, and with their rather formidable bill, adult ivory-bills probably have relatively few predators or competitors. In general these would likely be the same ones faced by the pileated woodpecker.[86] As with the pileated, the large cavity entrance of an ivory-bill nest or roost is big enough that an adept raccoon (*Procyon lotor*) could reach in. Rat snakes are almost certainly predators on eggs and nestlings, and a large rat snake might even take an adult. I once watched an encounter between a gray rat snake (*Elaphe obsoleta spiloides*) and a barn owl (*Tyto alba*) about the same size as an ivory-bill. The snake

Figure 1.4. James Tanner took this juvenile ivory-billed woodpecker from its nest in the Singer Tract in Louisiana on March 6, 1938. It was banded with US Biological Survey band #365-27264—the only ivory-billed woodpecker ever banded. The banded bird fledged about eleven days later. Although its juvenal plumage was similar to that of a female (no red on the head), by two and a half months after leaving the nest the bird had some scarlet feathers, indicating it was a male. *Courtesy James T. Tanner*

might have won had I not given aid to the bird.[87] Cooper's hawks (*Accipiter cooperii*), great horned owls (*Bubo virginianus*), and perhaps a few other raptors might be occasional predators on ivory-bills. Tanner recorded ivory-bill responses to Cooper's and red-shouldered hawks (*Buteo lineatus*) that were consistent with efforts to drive off a potential predator, but he also felt such predators were not a serious problem.[88] In Cuba, the Cuban crow (*Corvus nasicus*) has been thought to be a potential predator of ivory-bills, particularly the young.[89]

Many species—ranging from eastern bluebirds (*Sialia sialis*), to wood ducks (*Aix sponsa*), gray squirrels (*Sciurus carolinensis*), opossums (*Didelphis virginianus*), and honeybees (*Apis melifera*)—might take advantage of ivory-billed woodpecker cavities, but usually could not compete for one that was actively used by an ivory-bill. The honeybee might be an important exception, because it is not native to North America and thus ivory-bill experience with honeybees is relatively recent.

Tanner felt that the ivory-bill had no real competitors for its food, noting that ivory-bills and pileateds rarely seemed to notice one another.[90] Among Tanner's unpublished notes, shared with me by his wife Nancy, is an observation that he made one morning of two ivory-billed and two pileated woodpeckers feeding in the same tree. He did not record details of how they were feeding, what they were feeding on, how far apart they were, or any apparent recognition of either species by the other. The implication was that they were simply feeding together without interacting.

Tanner was observing ivory-bills that had, until shortly before his field work in the Singer Tract began, been living in an old-growth forest. Within a year, however, extensive cutting was underway. Under such conditions ivory-bills and pileateds had coexisted for millennia, each with its own specializations and with limited niche overlap. As a result of forest fragmentation and removal of dead wood and older trees, the habitat available to ivory-bills was reduced, perhaps narrowing or even eliminating foraging approaches that favored the ivory-bill over the pileated. This might well have forced the ivory-bill to feed on traditional pileated food. In such a situation the pileated probably had the advantage—a possible factor contributing to the decline of the ivory-bill.

Competition with the pileated woodpecker for cavity sites and for food resources is a potential problem that Lester Short has called serious.[91] Human impacts on southern forests may have thus favored pileated populations while negatively affecting ivory-bill populations. Short wrote that he was "pessimistic about chances for establishing a viable ivory-bill population in the presence of Pileated Woodpeckers."[92] On this issue I am ambivalent. I believe Tanner may have underestimated the competitive threat from pileateds in human-modified forests, but also that Short may have overestimated the threat. Certainly any discovery of ivory-bills should be followed by close scrutiny of the relationship of the two large woodpeckers.

By virtue of their residence in Cuba and in coastal plain ecosystems of the southeastern United States, ivory-billed woodpeckers are influenced by catastrophic weather events somewhat unique to the region. Hurricanes, lightning, and lightning-caused fires can be important in providing a continuous adequate supply of dead trees for nesting and roosting and as habitat for the beetle larvae the woodpeckers feed on. However, these events can also be lethal to the birds. The large, tall trees needed for cavity excavation would be especially vulnerable to lightning strike, and ivory-bills were likely occasionally killed when roost or nest trees were struck. With selective cutting of most older trees, the remaining tall trees might be not only a scarce commodity but also relatively more vulnerable to lightning strike or high winds.

Not so long ago, in the warm, moist forests of southeastern North America, it would have been possible to watch and study an ivory-bill family high up in a tree. Now we instead piece together day-to-day activities of the birds from accounts of deceased and aging observers, the fleeting unverified glimpses of hopeful birders, and stray threads from forensic ornithology. Perhaps we still have the chance to piece together how the ivory-billed woodpecker *lives,* rather than how it might have *lived.* Perhaps it lingers still in the remoteness that still typifies some of its range. If we do get the chance to see one fledge, to watch one find its food, to hear the double rap against a dying tree, we may be able to fill in much of the puzzle with the lost pieces.

The Land of the Ivory-Bill

[It] seeks the most towering trees of the forest; seeming particularly attached to those prodigious cypress swamps, whose crowded giant sons stretch their bare and blasted, or moss-hung, arms midway to the skies. In these almost inaccessible recesses, amid ruinous piles of impending timber, his trumpet-like note, and loud strokes, resound through the solitary, savage wilds, of which he seems the sole lord and inhabitant.

—Alexander Wilson, 1811, in *American Ornithology*

$\mathscr{W}$ith these words Wilson, recognized today as the "father of American ornithology," provided us with one of the earliest and most eloquent descriptions of ivory-billed woodpecker habitat as he found it twelve miles north of Wilmington, North Carolina, in the early nineteenth century.[1] More than a century later, in 1939, James Tanner added an important dimension to our understanding of ivory-billed habitat in a speech to the National Association of Audubon Societies: "The Ivory-bill has frequently been described as a dweller in dark and gloomy swamps, has been associated with muck and murk, has been called a melancholy bird, but it is not that at all . . . the Ivory-bill is a dweller of the tree tops and sunshine; it lives in the sun, not the shade."[2]

In the end we have lost, or almost lost, the ivory-billed woodpecker because we neglected to respect its home. So if we have a chance of keeping the ivory-bill—or understanding why we lost it—we must understand where it lived and what we did to that living space. It is not a simple story, because the ivory-bill is not a simple bird that lived in a single place. It is known from different habitats in the southeastern United States and Cuba, sharing its world with different ecological communities, and facing different challenges for survival.

Origins of the Ivory-Bill

The ivory-billed woodpecker is descended from a line of woodpeckers that seems to have originated in Latin America, where there are ten other *Campephilus* woodpeckers, including the largest woodpecker in the New World, the imperial woodpecker of Mexico. At some point scientists believe the habitats of the New World changed to favor the expansion of a South American *Campephilus* woodpecker north into what is now Mexico and the United States. Certainly the history of the Western Hemisphere is one of relatively frequent climate change, at least when measured over tens of thousands and millions of years. Such climate changes help explain the existence of one fossil of an ivory-bill relative, described as *Campephilus dalquesti* by Pierce Brodkorb. This fossil was discovered in Scurry County in central Texas, a relatively arid area now unsuitable for such woodpeckers.[3] The fossil has been dated to the Upper Pliocene, perhaps more than two million years ago.

Northern latitudes have experienced the ebb and flow of glaciers as major ice ages have occurred four times over the past two million years. An excess of rain in tropical areas matched the excess of snow and ice in the north. With the excess of rain, desert areas became forests, and fungi must have been rampant. The new forests were probably filled with wood-destroying insects and, as a consequence, woodpeckers. As glaciers retreated and climates became more arid, many forests of southwestern North America became deserts.

It is likely that during such a period of increased aridity in central Mexico, the ancestral population of ivory-billed and imperial woodpeckers was split into two populations. The incipient imperial woodpecker became isolated in the mountain forests of western Mexico and the incipient ivory-bill was forced eastward, ultimately to be isolated in forests of the southeastern United States and Cuba. The ivory-billed and imperial woodpeckers are closely related, differing slightly only in terms of size and plumage pattern, and they shared a common ancestor. The imperial lacks the ivory-bill's prominent white neck stripes and has black bristle-like feathers covering its nostrils, compared with the white bristles of the ivory-bill. The female imperial has an incredibly long, pointed, and up-curved black crest, whereas the female ivory-bill's black crest is not as long or up-curved. The differences are enough to convince most ornithologists that the two woodpeckers are separate species, but their similarities are such that scientists believe their divergence was in the geologically recent past—a divergence made possible when the relatively continuous forests of Mexico and the United States were disrupted during the warm, dry climates experienced between glacial periods.[4]

With the ancestral populations separated on either side of the arid center of Mexico, the eastern birds—the incipient ivory-bills—had two possible paths of dispersal to occupy their historical range. They could have gone northeast along the Texas coast to the southeastern United States and then from Florida to Cuba, or they could have gone from the Yucatan peninsula of Mexico to Cuba and from Cuba to Florida—or possibly they could have gone in both directions. During glacial times, when vast amounts of water were locked up as ice in northern regions, sea levels were lower by 300 feet or more, making Cuba much closer to both Florida and the Yucatan and dispersal between either and Cuba much more likely. I believe the most likely scenario is movement through Texas, along the American coast of the Gulf of Mexico, through peninsular Florida, and then over to Cuba during an ice age.

Another, albeit remote, possibility exists for the origin of the Cuban birds. When early explorers visited Cuba, they found that there was a lively trade going on between Native American people in Florida and Cuba, which included live birds.[5] The ivory-billed woodpecker was of great symbolic importance to indigenous people. Its heavy ivory-colored bill and the red feathers of males were often used as decoration and traded far outside the range of the species. Thus, it seems feasible that some of these birds could have been taken live from Florida to Cuba or from Cuba to Florida. Capture of a nesting pair would have been relatively easy, and later escape from captivity would seem quite possible. These birds stay together as they forage, a behavior that would have facilitated dispersal of pairs or family groups that might establish a new population. How many pairs would it have taken? Could they really have been introduced in this way to Cuba or the United States? Probably not, but continued studies of archeological evidence and molecular studies of tissues associated with specimens of the ivory-bill may clarify our understanding of the ivory-bill's population dynamics. The differences between North American and Cuban ivory-bills seem to be primarily in size, something that probably reflects adaptations to differences in the climates of the regions. Speaking from an evolutionary standpoint, bird size can change relatively quickly. Take, for example, the house sparrow (*Passer domesticus*), which since its introduction to North America from Europe a mere 150 years ago has exhibited changes of a similar magnitude.[6] Given that size is the major difference between Cuban and American ivory-bills, it is reasonable to believe the two populations were separated rather recently.

However ivory-bills became a part of the American and Cuban fauna, they were clearly adapted to life in the warm forests of both areas long before Columbus came to the New World. Early New World naturalists such as John James Audubon knew them well from the forests of the south.

Historic Distribution

The first good range map and detailed summary of the distribution of ivory-billed woodpeckers was that of Edwin M. Hasbrouck in 1891:

Prior to 1860 the Ivory-billed Woodpecker was distributed from Fort Macon, NC, along the coast as far west as the Brazos River in Texas, and extending towards the interior for an average distance of seventy-five miles; in the Mississippi Valley as far inland as central and western Missouri, southern Illinois, Indiana, and western Kentucky, together with a portion of Indian Territory.[7]

Although this fairly describes the general distribution of the species, more inland records and documentation of occurrence can be found in later summaries by Arthur Allen and Tanner; Tanner also elaborated on and clarified Hasbrouck's map.[8] The maximum historical extent of the species' distribution may be a bit less than that which was surmised by these authors. I have examined all these records and a number of other published records that were not included in the previous list, and I have examined specimens with locality data that have not been mentioned previously in the literature. I have looked at every verifiable record of which I am aware related to the places once home to ivory-bills and have included each on a revision of Tanner's map (see Figure 2.1) to make it as accurate as it can be with the information available. In studying this map, it is important to keep in mind the opportunistic nature of the ivory-bill. It would move into areas when there were high insect populations associated with tree mortality following floods, fires, or other catastrophes, and was only resident where such events occurred regularly. Because of this, the distribution of the ivory-bill within this range was not likely continuous, and population density likely varied greatly with habitat quality.

The distribution of ivory-billed woodpeckers as it is currently understood follows a clear pattern. The birds were found along major rivers where those rivers spread out across vast alluvial plains that were seasonally flooded and that supported extensive forest. They also ranged into adjacent upland forests. The broad distribution of ivory-bills in Florida seems to be an exception to this generalization. Although Florida rivers are generally small, adjacent lowlands and swamp areas associated with them form a mosaic that once encompassed much of the state.

Figure 2.1. The historic distribution of the ivory-billed woodpecker in the southeastern United States. Revised and redrawn from James Tanner, *The Ivory-Billed Woodpecker, National Audubon Society Research Report No. 1* (New York: National Audubon Society, 1942). *Courtesy National Audubon Society*

Dispersal patterns for ivory-bills no doubt varied with topography, thus also did their potential to survive catastrophic habitat losses. To the north, as riverine forests narrowed, dispersal of ivory-bills would have become more linear and their world more easily invaded by humans, with dispersal options limited. To the south, particularly in Florida, where the wetland mosaic remains extensive, dispersal options and links among populations would have been much greater.

In the late eighteenth century, Col. William Fleming observed ivory-bills in central Kentucky and had a servant shoot one, which he described in great detail. There would not have been extensive bottomland forest in the area, and thus these birds seem out of place.[9] Were the birds he found simply a wandering or dispersing pair? Or were ivory-bills once widely distributed in all the virgin forests of the southeast? Based on historical records such as Fleming's, localities from which specimens were collected, and our understanding of the ivory-bill's behavioral ecology, I suspect the ivory-bill was widely, though sparsely, distributed in virgin forests of the southeastern United States. Swamp forests were likely the heart of the ivory-bill's distribution, the habitats in which they were most abundant, and their final North American refuge from human hunting and clearing or fragmentation of the forests.

Ivory-Bill Habitats

Tanner wrote of the post-Columbian habitats, "All Ivory-bill records have been located in or very near swamps or Florida hammocks."[10] He went on to categorize the habitats as river bottoms, bottomlands of the Mississippi Delta, and the Florida region. What Tanner referred to as river bottoms were forests of the upper reaches of major southern rivers, those that flood for only a few months out of the year and are dominated by oak, sweetgum, ash, and other hardwoods. Upland southern pine–hardwood forests border these forests. As one ventures farther into the coastal plain, into areas with less physical relief, the bottomland forests, dominated by baldcypress and tupelo gum, are flooded most of the year. ivory-bills are known from both forest types, but Tanner felt that they favored the oak-sweetgum-ash forests; in other words, they preferred the drier sites to the baldcypress-tupelo-gum swamps.

The major differences between Tanner's river bottoms and the Mississippi Delta are the magnitude and diversity of forest types. The floodplains of other rivers are relatively narrow, whereas that of the Mississippi Delta combined with those of the adjacent waterways can be forty to eighty miles wide. In the Delta region, bottomland forest was once virtually continuous, although the tree species one encountered varied in response to minor changes in elevation. Tanner noted that records of ivory-billed woodpeckers were typically from the upper areas of the "first bottoms"—characterized by a forest mix of sweetgum, oak, ash, and elm. Pines historically were not a natural component of the Mississippi Delta forests, although they are relatively common today.

Tanner recognized different ivory-bill habitats in Florida. He included the ivory-bill habitat of north Florida in the river bottom category, but in the southern areas of the panhandle and in peninsular Florida, the limestone topography and sandy soils have generated a forest mosaic that is substantially different. From the southern areas of the panhandle southward, baldcypress was a dominant component of ivory-bill habitat. Other tree species associated with the ivory-bill in all habitats of the Florida region included black gum (*Nyssa sylvatica*), red maple, redbay (*Persea borbonia*), sweetbay (*Magnolia virginiana*), laurel oak (*Quercus laurifolia*), American elm (*Ulmus americana*), and cabbage palm. Unlike the other categories, this habitat more often occurs in patches as well as linearly along rivers, and interspersed among the baldcypress are occasional areas of pine flatwoods.

Tanner's three regions define a basic outline of the habitats important to ivory-bills, yet they do not tell the whole story. The forest had to be extensive. Tanner estimated ivory-bill density in three areas of extensive habitat. In the bottomland hardwood forest of the Singer Tract in northeast Louisiana he estimated there might have been seven pairs in 120 square miles—about one pair per seventeen square miles. In the California Swamp in north Florida, he estimated there had been about six pairs in sixty square miles—about one pair per ten square miles. And in the swamp forest along the Wacissa River in the Florida panhandle he estimated there had been about twelve pairs in about seventy-five square miles—about one pair per 6.25 square miles. The swamps occupied by ivory-bills were essentially virgin forest: most of them had never been cut or had experienced only minimal subsistence cutting.

Certainly the size of the trees favored by ivory-bills is important to their ability to excavate cavities in which to nest and roost. A bird this large has little use for small trees when it comes to making a home. It is true that many trees can grow to a trunk size that might be adequate for cavity excavation, but the smaller the tree, the more the birds are restricted to lower sites on the trunk—making their nests and roosts more accessible to predators. In the ancient trees used by these birds, a climbing predator might have had to search quite a while as it met gigantic limb after limb before getting to the cavity sheltering the woodpecker.

Older trees in older forests—those forests with ages measured in centuries rather than decades—have thicker bark with larger bark plates and deeper furrows between plates.[11] These older trees also have an increasing amount of dead wood that provides food and habitats for wood-boring beetles, whose populations can be larger and more stable as a result of increased habitat. This of course means a more steady, reliable food supply for the woodpeckers. Also of importance would have been the numbers of dead and dying trees, which provided habitat for the insects that the ivory-bills preyed on.

Little, if any, virgin ivory-billed woodpecker habitat remains today. Nevertheless, there may still be several areas of potentially suitable ivory-bill habitat. These are areas that include trees with a diverse age and species structure and that contain relict forests. Perhaps the weakest components of the possible habitats are the number of older and dead trees. With each passing year, however, the forests that are allowed to grow old improve as possible habitats for the ivory-bill.

The preponderance of habitat descriptions unquestionably link ivory-bills with mature riverine and swamp forests. However, I feel that ivory-billed woodpecker habitat preferences are much more catholic with regard to tree species but perhaps more restricted relative to other habitat qualities. Ample evidence of the species-wide habitat tolerance is demonstrated by their existence in the montane pine forests of Cuba, although it is possible they moved to these habitats after the lowland forests were cut.

Tanner recognized the greater diversity of habitats used by the ivory-billed woodpeckers in Florida and suggested that their ability to use diverse habitats was perhaps the reason why they were more common and more widely distributed there than they were elsewhere. Tanner was right, I believe, but not because of regional variation in the birds—rather because of regional variation in topography, habitats, and climate. The physical shape of habitat patches often influences the distribution of species that live there. Ivory-billed woodpecker populations along most major southern river systems would have been limited by the narrow corridorlike nature of the swamp forest habitat, although the Mississippi Delta forests included larger expanses of bottomland hardwood forest, and peninsular Florida presented more of a habitat mosaic.

The interrelationships among plants, animals, and fungi, and between living and physical factors in the environment are important. The trees provide nest sites and the foraging substrate on which the birds feed. They must be large enough for excavation of large cavities, and probably require the presence of heartwood-decaying fungi to facilitate excavation. The forests also must include an adequate number and dispersion of trees of an appropriate size and species to provide proper habitat for the beetles that the birds eat. And the beetles favored by ivory-bills are also specific in their needs, apparently inhabiting only recently dead trees.[12]

Losses of the virgin pine forests have not been linked previously to the decline of ivory-billed woodpeckers in North America, an oversight in my opinion.[13] Certainly losses of the virgin bottomland hardwood forests were of major and more recent significance to the ivory-bill. Yet there is significant evidence that pines were important to the woodpeckers. In Alabama, an 1891 record noted a nest of the species in a dead pine along the Buttahatchie River near Crump Springs. In 1909 in Florida, S. C. Graham described ivory-bills foraging on pine trees near the St. John's River.[14] Two decades later, A. H. Howell wrote of nesting ivory-bills, "Occasionally they choose a pine standing near the border of a swamp, or a palmetto, a bay, or a gum in a hammock." Regarding foraging sites Howell added, "The Ivory-bills feed largely in the open pine forests or in the palmetto hammocks near the swamps, and are said to be found frequently about burned tracts."[15]

Alexander Wilson certainly focused on the baldcypress swamps of North Carolina as ivory-bill habitat, but he also wrote,

Wherever he frequents, he leaves numerous monuments of his industry behind him. We there see enormous pine-trees, with cart loads of bark lying around their roots, and chips of the trunk itself in such quantities, as to suggest the idea that half a dozen axemen had been at work for the whole morning. The body of the tree is also disfigured with such numerous and so large excavations, that one can hardly conceive it possible for the whole to be the work of a woodpecker.[16]

The favored North Carolina habitat of the ivory-bill as described by Wilson thus seems to be a mosaic of baldcypress swamp and adjacent pine uplands, similar to the habitat in Florida. Wilson also documented the use of pines by ivory-bills in South Carolina, but pointed to the abundance of beetles in the trees. He described with loathing the outbreaks of what seemed to be southern pine beetles and suggested that blame for the forest's destruction was mistakenly attached to the ivory-bills:

> The diseased, infested with insects, and hastening to putrefaction, are his favorites; there the deadly, crawling enemy have formed a lodgment between the bark and tender wood, to drink up the very vital part of the tree. It is the ravages of these vermin, which the intelligent proprietor of the forest deplores as the sole perpetrators of the destruction of his timber. Would it be believed that the larvae of an insect, or fly, no larger than a grain of rice, should silently, and in one season, destroy some thousand acres of pine-trees, many of them from two to three feet in diameter, and a hundred and fifty feet high? Yet whoever passes along the high road from Georgetown to Charleston, in South Carolina, about twenty miles from the former place, can have striking and melancholy proofs of this fact. In some places, the whole woods, as far as you can see around you, are dead, stripped of the bark, their wintry-looking arms and bare trunks bleaching in the sun, and tumbling in ruins before every blast, presenting a frightful picture of desolation. And yet ignorance and prejudice stubbornly persist in directing their indignation against the bird now before us, the constant and mortal enemy of these very vermin.[17]

In Texas as well, records of ivory-bills associating with pines can be found. Vernon Bailey described observations of ivory-billed woodpeckers that were "so wild" that he got "only a long distance shot at one flying above the pine tops. The first two birds were seen calling and drumming near the tops of tall dead pines about six miles east of Tarkington Post Office in one of the many patches of pines killed by boring larvae."[18]

In an account from 1882, Henry Nehrling stated that in southeastern Texas he found the ivory-billed woodpecker "only in the large and dense pine forests in the northern part of Harris County and in Montgomery County far from human habitations."[19]

It appears, then, that ivory-billed woodpeckers will inhabit both hardwood forests of river bottoms and pine forests of higher elevations, particularly old-growth forests supporting healthy populations of beetles. They seemed to do best at the interface of these forest types, taking advantage of the resources of each.

Many of the beetles whose larvae feed on pines or hardwoods depend on a moist environment. The high humidity associated with swamp forest environments promotes fungal growth, which in turn weakens the wood, allowing more efficient use of the tree by many beetle larvae. Although all forests are humid at some times, the relative constancy of humidity and the increased thermal stability resulting from the high humidity in southern swamp forests are likely critical to the life cycle of the larger beetles, whose larvae require a particularly long time to develop.

While studying red-cockaded woodpeckers near Aiken, South Carolina, I had an opportunity to see and feel the dramatic influence that a difference in humidity can make within a forest. One of the stands that included nest and roost trees of red-cockaded woodpeckers was adjacent to a 1,000-acre lake that was a cooling pond for two nuclear reactors used to produce plutonium for the US nuclear weapons program. Water leaving the reactors and entering the pond was scalding hot near this stand of pines. The presence of the hot water had a dramatic influence on the local microclimate: heat and humidity were stifling, and the combination resulted in an abundance of fungi. Abundant also were woodpeckers of all species. Downed wood decayed rapidly, and pileated woodpeckers were particularly common, taking advantage of the abundant insects. Indeed, I have never seen pileated woodpeckers in similar numbers anywhere else.

Over time I have come to believe that the presence of pines and other softwood trees in or adjacent to a forested wetland—one that because of its height structure and canopy would hold humidity—are critical to ivory-bills. Under natural conditions these trees grew large and, when they died, the high humidity led to rapid decay and a supply of beetle larvae.

Warmer year-round temperatures enhance the rate of fungal growth and survival potential for the larger beetles, but higher temperatures alone do not generally create better woodpecker habitat. It is the combination of temperature and moisture that is the key. Thus, the humidity and temperature regimes of the available habitats in Florida and Louisiana are likely important in allowing the establishment of the high ivory-bill densities historically reported from those areas.

Photographs and descriptions of the habitat of ivory-billed woodpeckers suggest a range from an open canopy to a closed one. An enclosed canopy retards the drying effect of the sun, buffers the environment against temperature extremes, and promotes maintenance of high humidity. Thus, leaving a few dead trees in the middle of a clear-cut for woodpeckers, a common practice in forestry, results in trees baked hard and dry by the sun, leaving only the base optimum for beetles and fungi. A closed canopy can limit understory growth, but so too can factors such as flooding and fire—both prominent features that shape ivory-billed woodpecker habitats and provide conditions that support their food supply.

Within a forest, the trees are the dominant feature, but the understory vegetation often has a major effect on the animals living there. It is simply a fact that larger creatures need "elbow room." A dense understory or midstory would limit the flight of a bird as large as the ivory-bill, forcing it to occupy the more open areas. Also, a well-developed understory would make it more difficult to gain access to beetle-laden fallen trees.

Regional differences in climate and especially the incidence of severe weather also contribute to the nature and persistence of available habitats. For example, coastal regions of the Southeast are more vulnerable to hurricanes than are inland regions. Hurricane winds kill large numbers of trees, thus enhancing habitats for beetles and woodpeckers. In addition, the southeastern United States has one of the highest incidences of electrical storms in the world. Central Florida has more than ninety days per year on which there are electrical storms. Across north Florida this frequency drops to sixty days per year, and across the mid-South it averages only thirty days. A high incidence of electrical storms is important because lightning kills trees, thus providing food sources for the beetles and, subsequently, for the woodpeckers. In the pinelands especially, lightning plays a bigger role in that it starts more fires in this environment that in turn kill yet more trees, providing even more beetle habitat. However, even in the bottomlands large fires do occasionally happen. J. A. Putnam suggested that a serious fire season occurs on average about every five to eight years in the bottomland hardwood forests of the Mississippi Valley.[20] When a bottomland fire does occur, it typically moves rapidly across the floodplain floor, damaging or destroying shrubs, herbaceous cover, and trees younger than about ten years old. Such fires can injure larger trees, increasing their susceptibility to disease and insects, including beetles. Several researchers have noted that ivory-billed woodpeckers were found in recently burned areas, suggesting a link to fire events.

Delving as deeply as we have into ivory-billed woodpecker habitat reveals a complex set of interactions that results in the ideal situation for this large bird. Forest age, tree species composition, understory density, humidity, temperature, storm frequency, fire frequency, flood frequency, habitat shape, and other factors all have played significant roles in the ivory-bill's dance between survival and extinction. But in the past 200 years, this symphony of interactions between the ivory-bill and its environment has been drowned out and transformed by the guttural voices some call "progress."

CHAPTER 3

*Recognition and
Causes of Decline*

*The probabilities are that it will soon be extinct, so far as the United
States is concerned. It is too handsome not to tempt every collector,
taxidermist, and millinery establishment in the country, and he who
protests will be laughed at for his trouble.*

> —Charles C. Abbott, speaking of the ivory-billed woodpecker,
> 1895, in *The Birds about Us*

$\mathscr{A}$ physician and serious ornithologist in the waning years of the
nineteenth century Charles C. Abbott was not alone in his sentiments.[1]
By the late 1880s when Edwin Hasbrouck summarized the status of the
ivory-bill, it was clear that it was disappearing.[2] In the last years of the
nineteenth and the first years of the twentieth century, many observers
commented on the ivory-bill's imminent demise. Yet nothing was done
to help the species.

Philip Laurent of Philadelphia was the late-nineteenth-century
equivalent of an avid birder today—only then, avid birders tallied birds
shot and stuffed rather than ones seen. Few birds were protected by law,
and trade in bird specimens was much like trade in baseball cards today.

Laurent spent a good deal of time in the Gulf Hammock area of northwestern Florida, arguably the area of greatest ivory-billed woodpecker abundance in the late 1800s. He had collected five ivory-bills there in 1887, but in 1906 he noted, "Twenty-odd years ago the Carolina paroquet [*Conuropsis carolinensis*] was quite common in this neighborhood, and the same can be said of the ivory-billed woodpecker, but none of the former [has] been seen for twenty-odd years, while the latter is growing scarcer with each passing year."[3]

A conservation movement was beginning, but the major focus was on stopping the trade in feathers for use on ladies' hats. Another twenty years passed, and John Phillips wrote of the ivory-bill being "almost but not yet extinct."[4] He told of a few in Florida and suggested that scattered pairs might be found in south Alabama, Mississippi, Louisiana, and Texas. By the late 1930s, when James Tanner made extensive searches for ivory-billed woodpeckers throughout the southeast, the birds were certain to exist only in northeast Louisiana, but considered to possibly be hanging on in coastal South Carolina, Florida, and southeast Texas.

Historic Status and Patterns of Decline

The rarity of the ivory-billed woodpecker is a major issue in the discussion about its possible extinction and past conservation efforts. Some authors have argued that the ivory-bill has always been rare. Others, relying on such statements, then suggest that its extinction was to be expected because of its rarity. In turning to the earliest records, those from the first naturalists, we find the ivory-bill often described not as a rare find but rather as "abundant," "not uncommon," and "common." How do we deal with this discrepancy? My sense is that across the southeast, because of its size, its conspicuous features, its large home range, and its somewhat social nature, it was not a "common" bird but rather a "commonly seen" bird. The extent to which it appeared to be common must also be tempered with an understanding of human transportation in the early nineteenth century. Rivers were the major thoroughfares, and their surrounding bottomland forests had not yet been cut. On these rivers the early naturalists would have traveled through the heart of ivory-billed woodpecker habitat. As these magnificent birds flew back and forth across a river they would have been conspicuous. At best I believe they might have been locally common. Away from extensive old-growth forest ivory-bills would likely have been absent or rare.

What patterns can we see in the ivory-bill's decline? What caused its precipitous slide toward extinction? As with other endangered species, such as the red-cockaded woodpecker, the history of the distribution of ivory-billed woodpeckers is one of contraction from the periphery of its range following the arrival of Europeans.[5] This has been followed by continued losses in numbers, progressive fragmentation of remaining populations as original habitats have been altered, and extinction of local populations as a result of isolation, habitat destruction, and other natural and human-induced events.

Habitat Losses

Ask any ecologist. What is the single most important factor leading to the endangerment of species? The answer is clear: habitat destruction. In past eras, species such as the dodo (*Raphus cuculatus*), the great auk (*Pinguinus impennis*), and perhaps the Labrador duck (*Camptorhynchus labradorius*) have been pushed to oblivion by overhunting. Others have succumbed to such things as competition, predation, or habitat alteration by exotic species.

The case for the demise of the ivory-bill is complex, but habitat destruction, specifically cutting of the virgin forests, must lead the list of causative factors. Cutting was undertaken, of course, for many reasons, including clearing of forest to make way for homes, farms, towns, and cities, as well as for removal of wood and other forest resources for fuel, lumber, and other human needs.

Following the Civil War, vast acreage of southern lands reverted to the federal government. Recognizing that no local taxes were being paid on these lands, Alabama congressman Goldsmith Hewitt was successful in getting a bill passed that allowed for their sale.[6] They were quickly bought up, mostly by northern forest industries, at prices often averaging $1.25 per acre. The forest industry moved in and railroads were built to remove the virgin timber.[7]

One practice of the forest industry that might have temporarily benefited the ivory-billed woodpecker was that of girdling large trees to kill them during dry weather and returning to cut them down only when the water was high enough to float the logs out. Because ivory-bills are thought to favor the larvae of beetles that live in recently dead trees, this practice may have resulted in providing the birds with an added food supply. It may also, however, have lured them to their death, because bird collectors typically took advantage of the transportation and expertise of lumbermen.

Some virgin forest survived into the twentieth century, and conservation efforts saved some swamplands in the early years of the century.[8] But World War I brought increased demands for lumber. It was considered the patriotic thing to do to cut the forests (Figure 3.1). Harry Church Oberholser and Edgar Kincaid may not have been quite accurate when they suggested that, "By 1918, virtually 100 percent of the lowland virgin hardwood in the South had been chopped and sawed down."[9] But the ravages of reconstruction and World War I were devastating, and only isolated remnants of the original forests remained into the 1930s. Adding to these demands on the forests came flood control projects on the Mississippi, which allowed drainage of the Mississippi Delta and cutting of much of the remaining bottomland forest. World War II brought additional "patriotic cutting," and clearing of bottomland forests continued for croplands.

As mentioned, the reasons for cutting have ranged from need for wood products and agricultural land to need for flood control and better control of rivers for barge traffic. Every forest had its own tale. For example, when I visited the Academy of Natural Sciences in Philadelphia to examine ivory-billed woodpecker specimens there, I was surprised to find a folded and yellowed letter tucked under a specimen from Bolivar County, Mississippi. Curators at the museum knew nothing of the letter and were surprised at its presence in the specimen case. M. G. Vaiden of Rosedale, Mississippi,

TO LUMBERMEN!

For the support of our soldiers in France the Government must have wooden ships.

Without ships the war can not be won.

Without timbers ships cannot be built.

OUR COUNTRY LOOKS TO YOU!

Every swing of an axe, every cut of a saw may score as heavily as a shot fired from the trenches. Help our boys in France. With them win the war.

Make the world safe for Democracy.

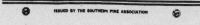
ISSUED BY THE SOUTHERN PINE ASSOCIATION

Figure 3.1. Placard placed in southern forests during World War I.

had written the letter in 1963 to James Bond of the Academy. In addition to providing more information about the specimen, the letter contributes to the knowledge of the role of the "exigencies of war" in the demise of the ivory-bill. Vaiden wrote, "p to the second World War there was another great estate composed of 12,341 acres of virgin timber located nine miles south of Rosedale, Mississippi, Bolivar County, and known as the Allan Grey Estate. It was reduced to zero during the War for the small PT boats used so effectively against the Japs in the South Pacific."[10]

Vaiden went on to provide insight relative to the understanding of the birds' needs and the lingering interest in collecting even such a rarity as the ivory-bill.

The only people I ever told of the ivory-billed woodpecker being in these great wood were George Lowery, Tom Burleigh, and Sam Ray. When the timber was sold to the government for federal use I thought these six pairs of ivory-billed woodpeckers would move to a small cypress brake of some eighty acres but they did not and so I lost an opportunity to take a bird for my collection. Where they disappeared to I have no idea.[11]

Ivory-billed woodpecker habitat loss sometimes was quick and dramatic, as with smaller areas of forest such as the Vaiden example. At other times, it stretched out for years. Vast areas were degraded, fragmented, and lost over such long periods of time that changes were little noticed from generation to generation. The losses began slowly and increased in severity and rate with human population growth and needs and with development of new forest industry technologies.

An example of such gradual ivory-billed woodpecker habitat loss is the development of forest industry in the Atchafalaya Basin of Louisiana. Cutting began in the Atchafalaya in the early 1700s when oaks were selectively removed. Although cutting of baldcypress for lumber was important then too, it was not until the mid-1800s that it became commercially significant.[12] During this time, two effective means of removing logs from swamp areas were developed: the overhead skidder and the pullboat.[13] By the 1890s, equipment was developed that allowed draining of some swamplands.[14] These developments, accompanied by a drastic decline in farming because of flooding, resulted in a tremendous growth in the lumber industry and increased selective removal of much of the virgin forest.

While Tanner was making his classic study of the ivory-billed woodpecker in Louisiana at the Singer Tract from 1937 to 1939, he could only watch as each day more habitat disappeared. He concluded that ivory-bills disappeared from most areas of their range shortly after logging began within their habitat, and that all other factors were inconsiderable in comparison. What Tanner witnessed, however, pales in the light of modern forest industry. Chainsaws had not yet been invented in the late 1930s, when all trees were cut by hand. Thus clear-cutting was virtually unknown then: it was simply too time-consuming to cut trees that could not be used. The pace of forest clearing accelerated by orders of magnitude with the use of chainsaws and modern heavy equipment from the late 1940s to the present. Little mature forest survived the 1940s, but the cutting has continued. Between 1937 and 1978, the extent of bottomland hardwood forest in the mid-South decreased by approximately 6.6 million acres.[15] With the widespread use of chainsaws beginning in about 1947, it became possible to clear-cut a large area in a relatively short time. Although by 1947, ivory-bills were already in a precarious position, the use of chainsaws made removal of trees from swamp areas more efficient and the potential for loss of remaining habitat was increased. Use of modern bulldozers and large-wheeled heavy equipment further facilitated habitat destruction. Although such mechanization fostered conversion of forest land to agricultural and other uses, it also paved the way for the introduction of "even-aged management"—clear-cutting and beginning a new forest of trees that are all the same age and of one commercial species. Such forests are typically harvested again before the trees reach old age, and neither the younger trees nor the lack of tree species diversity is conducive to providing even marginal habitat for a species such as the ivory-billed woodpecker.

Alteration of Fire Frequency and Intensity

Controling fire also changed the ivory-bill's world. The bird had evolved with the fire-climax ecosystems of the Southeast and took advantage of fire, finding insects attracted to recently killed trees and flying through forests kept open by fire. Several authors commented on ivory-bill use of such fire-dominated habitats. For example, in Florida in 1932, A. H. Howell noted that ivory-billed woodpeckers were attracted to recently burned areas as foraging sites, and Arthur Allen and Peter Paul Kellogg watched as a pair of ivory-billed woodpeckers left their roost and flew to a burned area to feed.[16]

Fire is a natural part of southern pine forest ecosystems as a result of a higher incidence of lightning in the Southeast than anywhere else in North America. Following the great cutting of the virgin forests in the late nineteenth and early twentieth centuries, however, there was so much logging debris left on the ground that wildfires were sometimes devastating.

Georgia forester Charles Newton Elliott suggested that catastrophic fires in the Okefenokee Swamp might have resulted in abandonment of an area by ivory-bills.[17] Fire came to be viewed as evil, and fire control and prevention became routine. However, forests "protected" from fire gradually changed in tree species composition, sometimes became too dense for such large birds to fly through, and likely provided less abundant food resources for the ivory-bill. The natural fire regime had been disrupted.

The positive ecological roles of fire were scarcely recognized until after the mid-twentieth century. Today we recognize that fire cannot play its natural roles in southeastern ecosystems as a result of modern roads that stop its spread, various control measures such as routine plowing of fire lanes between upland pines and bottomland hardwoods for prescribed burns, the use of prescribed burns at times other than when natural fire would have occurred, and the "Smokey the Bear" syndrome (suggesting that all fire is evil). Fire now almost certainly occurs in both upland and bottomland habitats with a lesser than natural frequency, and any remaining ivory-bills may be limited by dense understory, altered forest community structure, and reduced food resources.

Alteration of Natural Flooding Cycles

Equally or even more important than the altered frequency of fire in the ivory-bill's habitats is the alteration of water regimes. Much loss and alteration of swamp forest habitat can be linked to ditching and draining of swamps and the damming and construction of levees along rivers to protect cities and agricultural land, aid barge traffic, and provide hydroelectric power. Once swamp forests were drained or excluded from annual flooding, they could more easily be cleared and converted to agricultural land. Where the forests were not cleared, changes in tree mortality patterns and plant species composition, and development of a dense understory would almost certainly have had a negative impact on ivory-billed woodpeckers.

In Georgia, a movement for draining swamplands became well-established early in the twentieth century.[18] The swamps of the Altamaha River fell victim to such draining. Dams and levees along the

Mississippi and other southern rivers have had a profound impact on bottomland forests, grossly altering tree species composition and the whole nature of riverine forest ecosystems. And ivory-bills disappeared from their last South Carolina stronghold along the Santee River coincident with the upstream construction of the Santee–Cooper Hydroelectric Project in the late 1930s and impoundment in 1942.[19]

Changes in Habitat Structure and Diversity

The problems faced by the ivory-billed woodpecker only begin with the cutting of trees and the destruction of its habitat. As with all species, the interactions between this bird and its habitat are a complex web, and manipulation of individual components of that web can have cascading consequences.

Removal of trees from southeastern forests has largely been a matter of supply, demand, and efficiency of harvest. Oaks were typically the first to go among the hardwoods because of their use for heating and lumber. But as mills and harvesting techniques became more efficient and oaks less plentiful, other species were taken. Individual mills often concentrated on individual species until the supply was gone. The result has been continued alteration of tree species diversity, particularly among the older age classes. As noted, altered frequency of fire and flooding also altered tree species distribution patterns and abundance. To the extent that beetles the ivory-bills preyed on were host-specific, the birds' prey populations could have fluctuated dramatically.

Tanner noted that the ivory-billed woodpeckers in the Singer Tract foraged predominantly on sweetgums—the tree species that was being selectively cut from the Singer Tract in 1944.[20] There likely was an assemblage of beetle species that the birds relied on, each species with different habitat requirements. As the tree species, age, and structural diversity of the forest declined, so too would the diversity, abundance, and stability of the beetle community. Our understanding of basic ecology suggests that these changes in the birds' food supply could have resulted in their emigration, lowered fecundity, competition with other species, and local extinction.

Loss of bottomland forest habitat has resulted not only in reduction of habitat that could be used by ivory-bills for nesting and feeding, but also narrowing and sometimes complete loss of forested corridors along which the birds might disperse to any remaining suitable habitat. Dispersing birds would have had to traverse unsuitable habitat and would have perhaps been more vulnerable to predation. With fragmentation of the forests, small populations of ivory-billed woodpeckers became isolated from other such populations and thus more subject to stochastic factors and local extinction. Lack of gene flow was potentially a problem and is a correlate of habitat fragmentation and isolation of small populations. Allen and Kellogg suggested that the species' problems may have resulted from "sedentary habits resulting in inbreeding and weak young as soon as the colonies became isolated through commercialization and deforestation," or that breeding was impaired as a result of "lack of 'sex rhythm' for the same reason, resulting in infertile eggs."[21] But the immediate problems of habitat loss may have destroyed the species before genetic problems could have their effect. This would also lead to the disruption of the ivory-bill's social system.

Under natural conditions I doubt that competition and predation could have been significant factors in the ivory-bill's demise. James Greenway, in his review of endangered birds of the world, was unequivocal in his assessment of the ivory-bill's decline: "Neither those birds that compete with the ivorybill for food nor predators of any kind are to be held responsible."[22] However, with forest fragmentation and increases in edge habitats, we could expect increased populations of some potential ivory-billed woodpecker predators, such as great horned owls and red-tailed hawks (*Buteo jamaicensis*). Thus, it is plausible that they may have encountered increased predation as well as increased competition for dwindling resources.

Hunters and Collectors

Other major causes of ivory-bill decline were the hunter and collector. ivory-bills were hunted for various uses in Native American cultures, for food and as a curiosity by rural Americans, as prizes of economic value for amateur natural history collectors, and as documentation of the rarest of the rare for scientific collectors.

Some behavioral characteristics of ivory-bills and other woodpeckers make them particularly vulnerable to hunters. Pairs return to the same general area, often the same tree, to nest in consecutive years. Each individual has a roost cavity to which it returns at dusk each day. In addition, the young are in the nest so long and the parental instinct is so strong that once a nest was found, at least a pair of birds was assured.

The many uses of ivory-billed woodpeckers by Native Americans and their extensive trade in ivory-bill skins and bills would have had a negative impact that has been grossly underestimated. These were birds of great value that had come to have a special significance within Native American cultures. Under primeval conditions, killing an ivory-bill might have been a feat requiring considerable skill—although even then the birds could be counted on to return to a nest or roost hole, always giving the hunter another chance to kill it. With the introduction of guns to Native American culture, however, the rate and impact of killing of ivory-bills no doubt was greatly accelerated.

The extent to which ivory-bills were used as food by rural Americans is unclear. Laurent, a long-time resident of the Gulf Hammock area in Levy County Florida, in 1917 observed "most of the people living in the Hammock were acquainted with the ivory-bill, and the majority of the deer and turkey hunters had shot one or more of the birds."[23] He even went on to describe an old hunter who had shot many of them and "found them rather poor eating."

However, according to the early twentieth-century ornithologist Tom Burleigh, "Its flesh was said to be quite palatable, and since it was a large bird, it was consistently shot for food."[24] Arthur Wayne, a South Carolina lumber baron and serious amateur ornithologist, also noted that in Florida, ivory-bills were "shot for food, and the people—the crackers—consider them 'better than ducks.'"[25] In Cuba ivory-bills were also hunted for food.[26] Not that it answers the ivory-bill question, but

I can attest to the flavor of pileated woodpecker. While preparing a fresh road-killed pileated as a scientific specimen, I saved some of the breast muscle and grilled it. I cannot say that it tasted like chicken—rather, it was a bit gamy, maybe almost nutty, but not bad. Although the pileated had an acceptable flavor, a single bird provided little meat. I shudder to imagine how many ivory-bills it would take for a decent meal. Even if it was not a prized food item, at the end of a poor day of hunting, an ivory-bill might have been welcomed fare.

As with other game animals, hunters often found a variety of uses for the ivory-bills they shot. Their skull and bill were often used as curios or travel souvenirs. John James Audubon wrote that in the early 1800s travelers were "fond of possessing the upper part of the head and bill of the male . . . on a steamboat's reaching what we call a wooding place, the strangers were very apt to pay a quarter of a dollar for two or three heads of this Woodpecker."[27] Wayne included locals in this curio collector group as well when he noted that in Florida the bill of an ivory-billed woodpecker was "prized and many fall victims for that reason."[28]

Although travelers and locals might be satisfied with a head or a bill, serious bird collectors wanted a complete specimen. Ornithology during the latter half of the nineteenth century and early years of the twentieth century was dominated by efforts to document each local avifauna by collecting. Collections were the property of individuals, and hobbyists made many collections with no scientific aspirations. Even among scientists, there seemed to be competition to add the rarest of birds to one's own specimen cabinet.

Prominent ornithologists of the late nineteenth and early twentieth centuries were among those seeking ivory-bills. In the spring of 1890, ornithologists William Brewster and Frank M. Chapman explored the Suwannee River, specifically looking for ivory-billed woodpeckers and Carolina parakeets. On March 24, Chapman shot an ivory-bill. Instead of preparing it as a study skin or taxidermy mount, however, he prepared it as a skeleton—much to Brewster's disgust. Chapman did save feathers from his ivory-bill, however, and with some of them he made a tiny collage that included an ivory-billed woodpecker on a bald-cypress tree. He presented the artwork to Brewster as a Christmas present in 1890 (Figure 3.2). Another pillar of American ornithology of the era, Robert Ridgway, was photographed with one of the ivory-bills he collected in the Big Cypress Swamp in south Florida in 1898 (Figure 3.3).

Such was the order of the day that if one could not collect a specimen oneself, one purchased it. Placing an economic value on ivory-billed woodpecker specimens provided all the incentive that was needed for the birds to be hunted to the brink of extinction. In the June 1894 issue of the journal *The Oologist*, the ivory-billed woodpecker was listed among dozens of species for sale by F. H. Lattin and Company of Albion, New York. A "poor" specimen was available for $7, the "best" for $15. In the same year, Wayne published an advertisement on the back cover of *The Auk*, the journal of the American Ornithologists' Union, offering four pairs of ivory-bills for sale. He listed no prices. Charles K. Worthen advertised ivory-bill specimens in the May 1896 and other issues of *The Nidologist*, a bird magazine of the late 1800s with considerable emphasis on collecting.[29] His advertisements began, "Some bird skins can be secured elsewhere, but if you want the best and the rarest you should at once write Chas. K. Worthen, of Warsaw, Ill., who can give lowest rates on fine skins of Ivory-billed Woodpeckers."

Of course, when there was money to be made, there was also occasionally fraud. Pileated woodpecker eggs, for example, were sometimes advertised as ivory-bill eggs (Plate 1). *The Nidologist* dealt firmly with cheats—in print. In the June 1895 issue of Nidologist, a "Dr. Smith" is addressed in a four-stanza poem, part of which was the following:

O, Dr. Smith, of Malden, he
   Could lie and lie prodigiously!
Could lie in Texan, Illinois,
   Mexican, Piute, Iroquois—
Far greater than all kin or kith,
   Though he was simply Mr. Smith.[30]

Figure 3.2. The tiny collage of an ivory-billed woodpecker created with the feathers of an ivory-bill shot by Frank M. Chapman on March 24, 1890, along the Suwannee River in Florida and presented to William Brewster as a Christmas present. *Courtesy Ernst Mayr Library, Museum of Comparative Zoology, Harvard University*

The verse went on for three more stanzas. Apparently there were no worries about libel suits at that time. In the March 1896 issue the editor refers again to "the notorious Dr. Smith of Malden, of Ivory-billed Woodpecker fame," but unfortunately the reader is not informed of the nature of Dr. Smith's transgressions, although it was likely an effort to sell pileated woodpecker eggs as being those of an ivory-bill.[31]

Although several sets of ivory-billed woodpecker eggs may have been collected, only five sets of probably authentic ivory-billed woodpecker eggs have survived in museum collections: one set each at the Western Foundation for Vertebrate Zoology in California, University of Florida, and the University of Wisconsin at Stevens Point; and two sets

Figure 3.3. William Brewster and unidentified colleagues on March 24, 1890, along the Suwannee River in Florida. Brewster is holding the dead ivory-billed woodpecker that had just been shot by Frank M. Chapman (see Figure 3.2). This bird was made into a skeleton, much to the disappointment of Brewster, but the photo may reveal why Chapman chose to make it into a skeleton: Brewster is holding the bill in his right hand and enlargement seems to show that the tip of the lower bill had been broken by a pellet from the blast that killed it. It would have been an "imperfect" mounted specimen. *Courtesy Ernst Mayr Library, Museum of Comparative Zoology, Harvard University*

at the US National Museum of Natural History in Washington, DC (Plate 1).[32] In 1922, *The American Oologists' Exchange Price List of North American Bird Eggs* placed a value of $100 on an ivory-billed woodpecker egg, $750 for a California condor (*Gymnogyps californianus*) egg, but only $3 for the egg of a red-cockaded woodpecker.[33]

The last three decades of the nineteenth century were a vibrant time for the development of natural history museums in North America. The museum movement included both municipal and academic-based museums, and all were clamoring for specimens. Henry A. Ward of Rochester, New York, found his niche in this ferment of museum growth and development. He opened a business that provided specimens, cabinets, displays, and expertise to museums. He even sold ivory-bills—a specimen from Beaufort, South Carolina, in the collection at Muhlenberg College, Allentown, Pennsylvania, was purchased from Ward's Natural Science Establishment for $100, apparently in the 1930s. Ward is given considerable credit for the rapid growth of the museum movement.[34] As such it might be argued that he and other professional dealers in specimens were also increasing the value of rare specimens and hastening the demise of creatures such as the ivory-bill. When the Florida legislature passed a bill banning the killing of birds, the editor of *The Nidologist* lamented that no exception was provided for scientific collectors. When he discovered that the bill had only passed in the House and had failed in the Senate, he was much relieved.[35] A conservation ethic was growing, but it had not quite reached the collectors.

Passage of the Lacey Act in 1900 provided a legal tool that could help stem the trade in wild birds. The Lacey Act made it illegal under federal law for anyone to ship, receive, or possess wildlife if it had been taken illegally in one state and then shipped to another. In 1905, however, J. R. Jack killed ivory-billed woodpeckers in Florida and shipped them to dealers outside of the state. Unfortunately, he could not be prosecuted under the Lacey Act because it had been thought that the ivory-bills were extinct in Florida and they had not been given state protection. When he was released, he scoffed that he knew of more ivory-bills, saying that he already had orders for them. This time, however, agents from the National Audubon Society got him. They pretended to be dealers and got Jack to send them skins of northern cardinals (*Cardinalis cardinalis*) and northern bobwhite (*Colinus virginianus*), which were protected by Florida law. Jack was then convicted under the Lacey Act.[36]

The conservation ethic was alive in the scientific community—so long as it applied to someone else. In 1905, the American Ornithologists' Union Committee for the Protection of North American Birds reported on Jack, referring to him not by name but as a "commercial collector ... detected shipping skins of Ivory-billed Woodpeckers from [Florida] contrary to law."[37] They noted that he had been arrested and not yet tried and urged that "as the Ivory-bills are on the verge of extinction, he should be convicted as a warning to all persons who collect birds for commercial *rather than scientific purposes*" (emphasis mine). Apparently they still saw no problem with collecting ivory-bills for science.

In spite of the sanctioning of collection of ivory-bills a century ago, the numbers of ivory-bills in collections are relatively low. I have found records of only slightly more than 400 specimens, most of which were listed in Paul Hahn's *Where Is that Vanished Bird*.[38] Although we list these as "scientific specimens" today, most were not collected by scientists but rather by hobbyists. Fortunately many of these collections were ultimately donated or sold to scientific institutions. The peak period of collection of ivory-billed woodpeckers included the last two decades of the nineteenth century and the first decade of the twentieth. Some were collected well into the twentieth century (Figure 3.4).

The single largest number of specimens is about forty, housed at the Museum of Comparative Zoology at Harvard University. Ivory-bills, however, can be found in major museums around the world. I have examined specimens in museums across North America as well as in Austria and Cuba, and have obtained data from specimens in Russia, France, England, and Germany. Most museums—if they have any—have only one or two specimens. The Louisiana State University (LSU) has three study skins, which I examined in 1989. This is quite a contrast to the erroneous statement in one book that at the LSU museum there is "drawer after drawer in cabinet after cabinet containing ivory-bill skins stuffed with cotton, hundreds of preserved skins of males and females, juveniles and adults."[39]

Most existing specimens with locality data are from Florida, further evidence of the significance of the state to the species. Phillips said he had evidence that thirty-seven ivory-bills had been shot in the Oreilla Swamp near St. Marks, Florida, alone.[40] Those birds, he said, had been sold for $40 to $50 each—a price that suggests they might have been collected rather late.

We have learned much about the species from scientific specimens, and will continue to learn from them. But in combination with these other

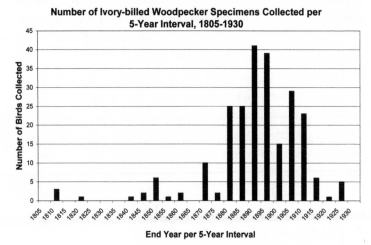

Figure 3.4. Numbers of ivory-billed woodpeckers collected per five-year period and now present in museum collections.

factors, collecting hastened the demise of the species. It is a sad testament that most museum specimens were shot during the nesting season, when they were most vulnerable, and include both members of a pair.

Following 1900, public sentiment rapidly turned against collecting of birds and their eggs, but many collectors continued—they were just more furtive in their activities. In 1927, for example, H. H. Bailey wrote that there had been ivory-billed woodpecker eggs collected since 1890, but they were not reported "due to the fact that adverse criticism appears as soon as some one publishes the taking of a rare record, thus eliminating that person from going on record again."[41]

For more than twenty years now I have been tracking down ivory-bill specimens, examining and measuring them, trying to learn as much as possible about them and the people who collected them. Most are lacking in data: we do not know who collected them, or when or where they were collected. Techniques used in taxidermy provided help with learning who prepared some specimens. A catalog number or taxidermy-shop advertisement stuck or written on the base of a mount helped me track details of others. Accession notes from some museums put values on some specimens and let us know that others were donated. Alfred Schmidt of the Taxidermy Shop of Tennessee, for example, donated two ivory-bills to the Museum of Natural History and Industrial Arts (now Pink Palace Museum) in Memphis, Tennessee, in October 1931.

A few ivory-bills remain in private collections to this day. Roger Tory Peterson (of Peterson field guide fame) had an ivory-billed woodpecker that had been collected by C. J. Maynard in November 1881. It had gone first to John D. Smith, then to Peterson. Maynard was a collector who made it his business. On one of the labels on Peterson's specimen is the statement: "Explorations in Florida by C.J. Maynard & Co., 36 Winter St, Boston, Massachusetts."

Although ivory-billed woodpeckers and other birds gained legal protection early in the twentieth century, and one taxidermist was arrested for trading in ivory-billed woodpeckers, collecting continued—sometimes with permits, sometimes without. In 1929, W. DeWitt Miller and colleagues noted the serious danger of extinction that was facing many North American birds and singled out the ivory-bill, noting, "Bird collectors will doubtless account for the few remaining individuals."[42]

In 1941, the humorist Will Cuppy offered advice to the ivory-billed woodpecker:

> Keep away from bird lovers, fellows, or you'll be standing on a little wooden pedestal with a label containing your full name in Latin: *Campephilus principalis*. People will be filing past admiring your glossy blue-black feathers, your white stripes and patches, your nasal plumes in front of lores, your bright red crest, and your beady yellow eyes. You'll be in the limelight, but you won't know it. I don't want to alarm you fellows, but there are only about twenty of you alive as I write these lines, and there are more than two hundred of you in American museums and in collections owned by Ivory-billed Woodpecker enthusiasts. Get it? [43]

As recently as 1943, Christy lamented that there were still "private collectors of hit-and-run morality."[44]

From personal experience on many occasions, I have learned that no particular excuse is needed by some individuals to kill a wild animal. Be it for curiosity or lack of a bigger, better target, or just plain boredom, any number of reasons undoubtedly resulted in killing of ivory-bills. As recently as the mid-1930s, former president of the National Audubon Society T. Gilbert Pearson told of Arkansas fishermen casually referring to using the flesh of pileated woodpeckers and other birds as catfish bait.[45]

Regardless of purpose, the ultimate factor—the bottom line—is people. Growing human populations and increased development of technology require more resources. Dissect this human factor and you find affectations of the human condition that have increased this race toward oblivion. War is certainly a major factor. The American Civil War, followed by World Wars I and II and more recent conflicts have taken a serious toll. During war, exploitation of resources is often justified as the "patriotic" thing to do—whether these resources are truly needed or not—and to oppose resource use in such a climate is viewed by many as tantamount to treason.

Bureaucracy and free enterprise are other major factors contributing to the loss of species. Bureaucracy can give us such wonderful laws as the Endangered Species Act of 1973. Free enterprise allows us to choose our course of action and to be different if we desire. But under a bureaucracy, change is exceedingly slow. That is good and bad. In the bad times some environmental damage is prevented by the buffering effect of bureaucracy; in the good times the buffering of bureaucracy can mean that free enterprise continues to cut the timber unabated, and important places such as the Singer Tract are gone before they can be protected.

CHAPTER 4

In the Time of the Tribes

A coffee grinder?

$\mathcal{I}$ could not believe it. For more than ten years I had been tracking down evidences of Native American use of ivory-billed woodpeckers. One afternoon I was at the Denver Museum of Natural History to examine ivory-billed woodpecker specimens and especially to look into a 1939 report of the bill of an ivory-bill that had been found in an Indian grave in Colorado. The presence of such bills in Indian graves far outside the range of the birds demonstrates their former value as a trade item among Native Americans. Both colonial naturalist Mark Catesby and John James Audubon wrote of such trade extending into Canada. But I had thought of it as something that occurred during colonial times or earlier. A. M. Bailey's report of the discovery of the bill in Colorado had left out details that now amazed me.[1] In addition to the ivory-billed and a pileated woodpecker bill, there had been a coffee grinder and some pieces of copper wire in the shallow grave with the two Indians. A coffee grinder! These were not *prehistoric* Indians but ones who enjoyed their morning cup of java. This vestige of an ivory-bill mixed literally with pieces of wire and a coffee grinder spoke of the ivory-bill's place in the culture of America's first people.

We may never know all that the ivory-bill meant to Native Americans, but we can get a sense by examining the traces left behind. Trade in ivory-billed woodpeckers among Native Americans had been going on for centuries and may have continued through the nineteenth century. Bills and other parts of ivory-billed woodpeckers show up regularly in graves and middens of Native Americans throughout eastern North America, both inside and outside the historical range of the species.

Archaeological records of birds are typically based on individual bones associated with the refuse tossed aside by an ancient civilization. Imagine the leg, wing, and breast bones of chickens left behind at a picnic area: scavengers would make short work of most of them, but occasionally one might be dropped into a fire that would consume flesh and oils that might attract a scavenger, or perhaps they would be buried in an effort to avoid attracting such scavengers. Some bones or pieces of bones took on a utilitarian value in crafting tools, musical instruments, or decorative items. Some, such as the bill of an ivory-billed woodpecker, seem to have had great symbolic value. Such items of value were often buried with their owner or a person of importance. Identification of birds from individual bones and interpretation of the circumstance under which the bones came to be at a particular site are often difficult. Paul Parmalee argued that the presence of ivory-billed woodpecker bills associated with Native American archeological sites suggests they might have been trade items from elsewhere but that bones not associated with the skull, which seem to have not had cultural significance, might more likely indicate local occurrence of the birds.[2]

To provide examples of the nature of such archeological evidence and its distribution inside and outside of the historical range of the ivory-bill, consider the following. In northwest Georgia, north of the historical range of the species, ivory-bill bones are known from the Etowah Indian mounds along the Etowah River.[3] In Illinois, at Cahokia, just across the Mississippi River from St. Louis, Missouri, a leg bone from an ivory-bill was found in a midden dated to have originated between 1000 and 1200 AD.[4] At a Sauk–Fox Native American cemetery in Rock Island County in northern Illinois, Parmalee found the lower bill of two ivory-bills.[5]

Far north and west of the known range of the ivory-bill at the site of the Big Village of the Omaha on the Missouri River in eastern Nebraska, archeologists found five ivory-billed woodpecker heads associated with Indian burials.[6] This is a site that was occupied from about 1780 to 1840 and was visited and described by Lewis and Clark. In each case, the ivory-bill head included only portions of the cranium and upper bill, and the head was placed beneath the right side of the individual buried. Several of the heads had red pigment associated with them. Similar parts of the skull and bill of three pileated woodpeckers, four bald eagles, a common raven (*Corvus corax*), two mallards (*Anas platyrhynchos*), and two common loons (*Gavia immer*) were associated with other burials.

Just beyond the historical western limits of the ivory-bill in Limestone County, east of Waco in east-central Texas, a wing bone tentatively identified as that of an ivory-bill was found at an archeological site dating from the late eighteenth century.[7]

In Scioto County in southern Ohio, Alexander Wetmore found bones of ivory-billed woodpeckers in Native American archeological sites that seemed to date from the fifteenth or sixteenth century.[8] Wetmore assumed that the ivory-bill was a local resident then, although no historical records of the species are known for the state. Additional prehistoric ivory-bill records come from Indian middens in Ross and Muskingum Counties in east-central Ohio.[9] Radiocarbon dating of material from the Muskingum site suggests that ivory-bills might have been in the area between the twelfth and fifteenth centuries.

In West Virginia, partial bills of ivory-billed woodpeckers were found in the Fairchance Indian burial mound in Marshall County, and a partial skull was found at the Buffalo burial site in Putnam County.[10] All of these artifacts might have been acquired by the Indians through trade rather than obtained locally. George Hall felt that the partial skull,

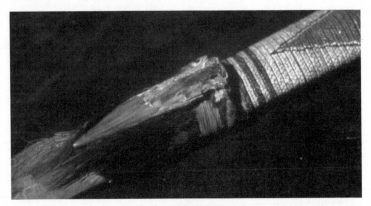

Figure 4.1. Detail of the stem of an Ioway war pipe showing the attachment of the bill and scalp of a male ivory-billed woodpecker. I suggest that the red geometric designs on the woven grass ornamentation on the pipe may represent additional bills. They, like the real ivory-bill bill, are oriented toward the smoker. *Courtesy Jerome A. Jackson*

which had no bill, suggested a local record, although I think it would be quite possible that the purchaser had simply removed the bill, possibly for use on a pipe (see Figure 4.1).[11]

Significance to Native Americans

Without question, woodpeckers were symbolically important to Native Americans long before the arrival of Europeans and perhaps into the twentieth century. To attempt to understand the nature of the ivory-bill's symbolism, I have studied historical references of ivory-bill use by Native Americans, specimens among cultural artifact collections, and the prehistoric use of ivory-bills as suggested from the archeological record. Interpretations of the information available concerning use of the ivory-bill is fraught with problems. Early references—and even later ones—are subject to considerable cultural bias by the interpreter, including me.

There's a Tradition amongst them that the Tongue of one of these woodpeckers dryed will make the teeth drop out if picked therewith, and cure the toothache (tho' I believe little of it and look on it as ridiculous) yet I thought fit to hint as much that others may try; for some such old Stories refer to some peculiar Virtues thos not to all that is said of them.[12]

So wrote Rev. John Clayton in 1693 as an annotation to his list of the birds he had seen in Virginia. Although he had noted that there were many different woodpeckers in Virginia, his descriptions suggest that he had not seen an ivory-bill. This belief about woodpecker tongues, however, seems to have been widespread among Native Americans and would certainly have included the ivory-bill.

It seems that woodpeckers were important to many, perhaps most, groups of Native Americans in several ways: putative medicinal values, food, symbolism for successful warfare, hunting or power, and perhaps others. Among woodpeckers, however, the ivory-bill seems to have been particularly important.

It was the colonial naturalist Catesby who, in 1731, told us that the bills of ivory-billed woodpeckers "are much valued by Canada Indians, who make coronets of them for their princes and great warriors by fixing them round a wreath, with their points outward. Northern Indians having none, they buy them from southern Indians at two or three buckskins a bill."[13]

Certainly the bills and scalps of the ivory-bill and scalps of other woodpeckers were a common currency among native Americans. They had value. They had cultural significance. What that significance was I believe may have varied from region to region and tribe to tribe. Archeologist M. R. Harrington noted that parts of woodpeckers were commonly included in personal amulets of the Sac and Fox tribes, as well as in hunting, medicine, and war bundles.[14] He suggested that a medicine pouch containing a dried ivory-bill or its head was believed to bestow to the individual the bird's powers to seek out and capture hidden prey.

Woodpeckers are very skilful in finding their prey, even when well concealed—a quality which made their skins desirable as amulets, for either hunting or war, as by this aid the Indian believed he could locate his enemy or even a deer, just as the bird can locate the larvae hidden away beneath the bark of a tree. Besides this, as one Indian put it, "The bird can peck a great hole in a tree in a short time; the warrior who wore the skin could do the same thing—it did not take him long to make a great hole in the enemy."[15]

In 1811, Alexander Wilson, certainly aware of Catesby's comments, also noted the importance of the ivory-bill to Native Americans. His experience was primarily with the birds in North and South Carolina, but his knowledge of Native American use of ivory-bills was also likely influenced by John Abbot of Georgia. Wilson wrote of southern Native Americans using them as amulets and ornaments:

An Indian believes that the head, skin, or even feathers of certain birds confer on the wearer all the virtues or excellencies of those birds. Thus I have seen a coat made of the skins, heads and claws of the Raven; caps stuck round with heads of Butcher-birds, Hawks and Eagles; and as the disposition and courage of the Ivory-billed Woodpecker are well known to the savages, no wonder they should attach great value to it, having both beauty, and, in their estimation, distinguished merit to recommend it.[16]

Audubon also weighed in with comments on Native American use of ivory-bills: "It is a beautiful bird, and its rich scalp attached to the upper mandible forms an ornament for the war-dress of most of our Indians. . . . I have seen entire belts of Indian Chiefs closely ornamented with the tufts and bills of this species, and have observed that a great value is frequently put upon them."[17]

Association with Pipes and Medicine Bundles

In addition to ornamentation of war dress and amulet use, Native Americans used the bills and scalp of the ivory-bill on ceremonial war pipes. Edward D. Crabb noted that a "peace pipe" of the Iowa (a.k.a. Ioway) Indians of Oklahoma in the Milwaukee Public Museum was ornamented with six bills and crests of the ivory-billed woodpecker.[18] The Ioway, a tribe related to the Sioux, occupied much of the central Midwest, from what is today Iowa, south to southern Missouri.[19] Ivory-billed woodpeckers would have occurred only in the southern part of the tribe's homeland.

When I visited the Milwaukee Public Museum, I found that there were actually four Ioway pipes that had ivory-billed woodpecker bills attached to them. On examining and discussing them with curators at the museum, I learned that these were *not* peace pipes at all but rather "war pipes." The stems of peace pipes, it seems, were round in cross-section, whereas war pipes were flattened. The bills of the birds are attached to the pipe stem; the stone bowl of the pipe is separate and attached for smoking.

The pipe stems in the Milwaukee Museum are each nearly four feet long, made of wood, wrapped with plant fiber and trimmed with a variety of materials.[20] Each pipe stem is unique, but the materials used included the ivory-billed woodpecker bills and scalps and such additional things as horse hair dyed red, porcupine quills, ribbon, and sinew. The fiber covering the pipe stems included geometric designs that to me seemed to represent additional bills. The designs were bill-shaped (triangular) and oriented toward the smoker, as were the woodpecker bills (refer back to Figure 4.1).

In preparing a bill and scalp to be placed on a pipe, the head was severed from the body, the skin of the head pulled back over the bill, then the skull and lower jaw were removed, apparently just barely leaving a connection between the skin and the upper bill. The skin was then pulled back and the prepared skin with the upper bill appears to have been "pulled" onto the pipe stem like a sleeve. When in position, the bill was folded back so that it pointed toward the smoker and the bill and scalp fixed in position with sinew and plant fiber such that the bottom of the upper bill was exposed against the red of the scalp.

The bills and scalps were arranged along the upper surface of the pipe stems. All of the crests and bills adorning these pipes were those of male ivory-bills, thus it seems that a particular value was placed on the red crest. It seems likely to me that the red crest on the head might have taken on a symbolic meaning—perhaps suggesting a bloody head, an omen for success at war.

In contrast to the current belief that these were war pipes, A. Skinner, who spoke with tribal members in the early twentieth century, reported that the position of the bill over the red feathers was "to hold down the crest, for these birds erect their topknots when angry, and this is a peace pipe, and hence removed from wrath of all kinds."[21] It is true that an "angry" woodpecker would raise its crest and that each bill on each of the pipes I have examined is folded back over the crest.

The Omaha Indians also used the scalps and bills of ivory-bills on their sacred tribal pipes. A. C. Fletcher and F. LaFlesche suggested that a pipe with seven heads symbolized the council of seven chiefs, whereas one with a single head represented the unity of authority of the chiefs.[22]

Charles C. Willoughby, in his *Notes on the History and Symbolism of the Muskhageans and the People of Etowah* (a site in Georgia) mentioned the presence of two additional pipes with ivory-billed woodpecker bills and scalps on them as being in the Peabody Museum at Cambridge. Willoughby described them:

> . . . on the upper side of the stems . . . are fastened several upper mandibles . . . with scalp and crest attached. The mandibles are turned backward over the scalp that is tightly bound to the pipe stem. One of the pipes is a war calumet with the pendent attached of eagle feathers stained red. The other is a peace calumet with eagle feathers unstained. The stem and appendages of the calumet all had their symbolic meanings, and together formed one of the most highly ritualistic emblems known.[23]

Thus, similar pipes appear to have been used in the Southeast and in the Great Plains. Ivory-billed woodpecker bills and scalps were also found on a Native American pipe brought back from the Lewis and Clark expedition, probably also from the Great Plains.[24]

My suggestion of the significance of a seemingly disproportionate value placed by Native Americans on the bills and crests of male ivory-billed woodpeckers over those of females is also supported by the symbolic use of red feathers in war bonnets of the Dakotas (known also as Sioux). Stanley Vestal suggested these red feathers were used to symbolize wounds received in battle.[25] Similarly, among the Ojibway of the Great Lakes Region and southern Canada, the red feathers from a red-headed woodpecker's head were said to be a symbol of "valor, and were chosen to ornament the warrior's pipes."[26]

A photo of a medicine bundle of the Potawatomi Indians of Kansas shows that the bundle includes the partially mummified skins of at least three ivory-billed woodpeckers. These included at least one male and female; only the bill of a third bird is visible.[27] The wings of at least the male were bound with red cloth. I had no luck in locating these specimens.

In a reference that almost certainly refers to the ivory-billed woodpecker, an English translation of the French ornithologist Jules Michelet's book *L'Oiseau*, published in 1869, speaks of the importance of woodpeckers to Native Americans: "They are partial to wearing the head of one which they name 'the wiry-billed woodpecker,' and believe that his ardour and courage will pass into them."[28]

This early reference may help explain why the bills of the ivory-bills on the pipe stems face toward the smoker, allowing the smoker to breathe the breath of the great bird.

Other Ornithological Evidence for the Use
of Woodpeckers by Native Americans

Almost certainly the various Native American tribes of the Southeast had specific names for the ivory-bill, but their languages were not written until European influence, and then we find little interest in such a subject as the ivory-bill. Among the names that come down to us is the Seminole name *Titkka* for the ivory-bill recorded by Walter Hoxie.[29] The word *Wahzhi gapa*, meaning literally "bird head," was used by the Omaha Indians to refer to both the ivory-bill and the pileated woodpecker.[30] The Omaha apparently obtained the head of both species with the attached upper bill through trading with southeastern tribes.

Quatotomomi, according to Louis Jean Pierre Vieillot in 1807, was the name used for the ivory-billed woodpecker by natives in Mexico.[31] Because the ivory-bill does not occur in Mexico, we are left wondering. Is this a name used in Mexico for the imperial woodpecker, which was not described for science until decades later? Or is this a name from somewhere else within the ivory-bill's range? One possibility is that all of Texas was once considered part of Mexico, and perhaps Vieillot was referring to ivory-bills and Native Americans in east Texas, where they were known.

In addition to the reported use of woodpecker tongues to deal with toothache, woodpeckers have also been suggested as useful to the northern Shoshone in curing venereal disease.[32] In the Pacific Northwest, Klamath and Shasta shamans wore woodpecker scalps on headgear, and they were worn as part of the attire used by dancers participating in Yurok and Hupi jumping dances among Indians in California.[33]

Archaeological Evidence for Native American Use of Ivory-Bills

Native American breast plates made of large clam shells on which a circular design of four linked, crested woodpecker heads had been etched (Figure 4.2) are known from Kentucky, Tennessee, Mississippi, and Alabama—essentially the drainages of the Cumberland and Tennessee Rivers.[34] Such breast plates are known as *gorgets,* and the particular style that includes the woodpeckers is known as the *Cox Mound* style. These gorgets are linked with the Mississippian period of prehistory and believed to have been produced from about 1000 to about 1700 AD, with most produced after 1400 and before 1650 AD.[35] Jeffrey P. Brain and Philip Philliips, G. E. Lankford, and William H. Holmes illustrated several of these woodpecker gorgets showing four crested woodpeckers on each.[36] Being stylized, these could represent ivory-billed or pileated woodpeckers or both. Holmes linked these gorgets to mound-building cultures of the region. Lankford did not identify the culture of origin for the designs but referred to them as "wind gorgets," suggesting

Figure 4.2. Shell gorget with woodpecker motif.
Courtesy Jerome A. Jackson

that the four woodpeckers represent winds from the cardinal directions, but also acknowledging that we do not really know what they mean.

In 1998, while traveling across Tennessee, I stopped at an antique flea market and found one of these woodpecker motif shell gorgets for sale for $300. I was tempted by it, but concerned about its origin. In later talking to an anthropologist colleague about it, I learned that there is a brisk trade in counterfeit gorgets and that the one I found was likely of recent origin.

Among the Calusa Indians of Florida, a culture present when the first Europeans arrived but that disappeared by the nineteenth century, an ornate woodpecker motif that could represent the ivory-bill or the pileated woodpecker (or both) was found painted on wood and engraved on metal ornaments.[37] The metal ornaments are particularly intriguing because (1) we do not know what they were used for, and (2) we can be sure that they originated after the arrival of Europeans, because the Calusa had no access to metal. Such ornaments were created from metal items obtained through European contact. One suggestion is that they may have been used in the hair. R. J. Wheeler considered a religious tie, suggesting that the Calusa believed that following death the soul (or one of the souls) entered the body of another creature.[38] Nine metal-crested woodpeckers found are fairly similar and are known from sites ranging from southwest Florida to the Florida panhandle. The woodpecker painted on wood came from Marco Island in southwest Florida.

A Continuing Influence on Native American Culture

In November 1995, the Apalachicola Creek Indians, descendants of those who had not been removed from Florida, met to install a new "Mekko"—their king. During the ceremony, by tradition, into Mekko's right hand was placed the wing of a wild turkey (*Meleagris gallopavo*), which he held in a manner to show off the plumage. Observers at the ceremony noted that the last time such a ceremony was held, in 1923, an ivory-billed woodpecker skin had been used instead of the turkey wing.[39] The explanation for change was that

> Sometime during the early 1960s, a person representing himself as a game and wildlife agent for the federal government removed the bird skin from our possession. It was said that it would be returned after study and comparison. All now feel we were conned and the person did not represent a legitimate authority. A Pileated Woodpecker is an equal and acceptable substitute but apparently not to be made available to us, even as a loaned item. Should either bird become available, the entire rites would be immediately repeated.[40]

For many reasons the ivory-billed woodpecker seems to have been important to Native Americans both within its southeastern range and among tribes that traded with them, sometimes hundreds of miles away. Indeed, the uniqueness of the ivory-bill and its presence in cultural materials provides strong evidence of the extent of commerce among Native Americans. The ivory-bill's large size, the striking crimson and black crest of the male, its spear-like tongue, and its massive ivory-colored bill all contributed to its mystique. This bird obviously symbolized power and seems to have played somewhat of a spiritual role in Native American lives. The ivory-bill's disappearance coincided with dramatic cultural losses and changes as the melting pot of Euro-America came to a boil during the late nineteenth century. Descendants of those tribes in search of their spiritual past and cultural identity look to this will-o-the-wisp even as birders and scientists grasp fervently at any hint of its survival.

Part Two
The Searchers

CHAPTER 5

*The Discoveries of
Early Naturalists*

*"...when another Age is come, the Ingenious then in being may stand
upon the Shoulders of those that went before them, adding their own
Experiments to what was delivered down to them by their Predecessors,
and then there will be something towards a complete Natural History...."*

—John Lawson, 1709, *A New Voyage to Carolina*

*J*ohn Lawson was a young man who came to the Carolinas in 1700
from England because someone suggested it was the best place to go.
He obtained employment as a surveyor and traveled overland and by
canoe from Charleston through South Carolina and North Carolina,
ending up near what is today Washington, DC. He traveled by canoe up
the Santee and Cooper Rivers, marveling at the size of the trees in the
adjacent forests. This was ivory-bill country, a place where ivory-bills
existed into the twentieth century. He likely saw the birds, although he
may not have recorded them. During his travels he recorded much
about the Native Americans and natural history he observed. Among
the birds he saw, he noted that he had seen four kinds of woodpeckers.
He had no names for them, but described one as being "as big as a

Pigeon, being of a dark brown Colour, with a white cross on his back, his Eyes circled with white, and on his Head stands a Tuft of beautiful Scarlet Feathers."[1] Alas, there were no field guides, checklists, or even binoculars in the early 1700s. It is difficult to know what bird Lawson is describing, but the presence of the cross on the back along with a scarlet crest suggests an ivory-billed woodpecker. On the other hand, ivory-bills are larger than a pigeon and are truly black, not dark brown, and their eyes are not circled with white—although ivory-bills have a pale yellow iris. It is thus pretty tough to credit Lawson with the earliest written description of an ivory-bill. Our earliest good description and illustration of the ivory-bill comes from Mark Catesby.

Mark Catesby

Catesby was born in England in about 1682.[2] All we know of his early interests in natural history comes from the preface to his *Natural History of the Carolinas*, where he noted that his interest in nature was suppressed because his family lived too far from London and thus he wasn't exposed to the "Center of all Science."

In any case, by 1712 Catesby made his first trip to the New World, made possible by his father leaving him a good inheritance and his sister's living in Virginia with her husband, Virginia's secretary of state, William Cocke. It seems likely that Catesby's first trip to America was partly subsidized by patrons for whom he was supposed to collect plant specimens. He did take a few plants back to England from that trip, and perhaps a few of his paintings, but we have no record of any concerted natural history studies. On his return to England in 1719, Catesby noted that he regretted not having spent more time studying natural history in America.[3] He was able to use the plants he brought back and his few drawings to garner support from wealthy patrons for a return trip to collect additional natural history specimens for preparation of a book on the natural history of British America. In part this new enthusiasm was stimulated by the friendship and support of Sir Hans Sloane, one of the founders of the British Museum.

In April 1722, Catesby returned to America. During the first three years of this visit, he traveled in South Carolina, Georgia, and Florida. Then, before returning to England, he visited the Bahamas. Unfortunately Catesby did not record his travels or the locations where he collected specimens. In 1726 he returned to England and began writing *The Natural History of Carolina, Florida, and the Bahama Islands*. His first volume, published in 1731, included 200 of his hand-colored etchings, mostly of drawings of birds with plants that characterize their habitats. The second volume came out in 1743 and includes a much broader range of animals. We know little about the influences on Catesby's work, the origins of the names he used, or of his personal experiences with the creatures he portrayed. When published in England, however, his work was an immediate success.

By the 1700s, the European avifauna had been diminished, and although Catesby's interest in birds seems to have been secondary to his interest in plants, he was impressed by the diversity of birds in North America. Catesby explained his focus on birds:

> There being a greater Variety of the feather'd Kind than of any other Animals (at least to be come at) and excelling in the Beauty of their Colours, besides having oftenest relation to the Plants on which they feed and frequent; I was induced chiefly (so far as I could) to compleat an Account of them, rather than to describe promiscuously, Insects and other Animals; by which Method I believe very few Birds have escaped my Knowledge, except some Water Fowl and some of those which frequent the Sea.[4]

Ultimately, Catesby is best known for his work with birds.

We do not know where or when it was that Catesby found ivory-bills, but we do know of travels that would have taken him into good ivory-bill country. In 1722 to 1723 he spent a year in coastal South Carolina near Charleston, not far from the great swamps of the Santee and Cooper Rivers, places from which ivory-bills were known. In the summer of 1723, he visited the middle Savannah River area just south of Augusta, Georgia, staying at Fort Moore and roaming the nearby

Figure 5.1. The "Largest White-Bill Wood-Pecker" of Mark Catesby. *Courtesy Albert M. Greenfield Digital Imaging Center for Collections, Academy of Natural Sciences, Philadelphia*

countryside in search of birds and plants.[5] This area of the Savannah River Swamp is also an area where Alexander Wilson found ivory-bills.

In 1731, Catesby provided the first detailed, published description and illustration of the ivory-bill (Figure 5.1).[6] He gave the bird the common name of "Largest White-bill Wood-Pecker," but added that the bill was "white as ivory," most likely leading to the name we know today. He provided us with the first weight of an ivory-bill ("twenty ounces"), the first indication of their food habits, and a graphic description of their ability to excavate a nest by noting that "in an hour or two of time they will raise a bushel of chips." Catesby also provided the first intimation of the importance of ivory-bills to Native Americans and of trade in ivory-bills extending to Canada, far outside the species' range.

Col. William Fleming

Although Catesby provided a graphic illustration of the ivory-bill, a much more detailed description came from an individual unknown as a colonial naturalist but rather as a capable individual well-versed in science. Col. William Fleming, a Scottish immigrant and surgeon who joined the Virginia Commission, was sent to Kentucky to settle land claims.[7] While in what is today Lincoln County in south-central Kentucky, on March 7, 1780, Fleming saw two birds he did not recognize. He described the colors, pattern, and dimensions of the bird in such great detail that there is no doubt as to its identity:

> Rode up to St. Asaphs from Col. Bowmans, I observed a species of the woodpecker which I had not met with before, the cock and the hen, they are larger than the large brown [undoubtedly referring to the the pileated woodpecker], the cock had a bright red head with remarkably long tuft of feathers on the Crown so that it may be cald the Peacock Woodpecker the body & wings white & black, the hen darker colored the bills of both a great length & white.[8]

As with other early naturalists, the tool of greatest use in bird study was the gun. There were no binoculars and no field guides. Identification was often based on comparison of a bird in hand with lengthy published descriptions—and often descriptions were not available. Usually the best a curious naturalist could do was to shoot the bird, describe it in written detail, and possibly sketch it. Preservation of specimens was possible, but often not practical because of insects and other pests that destroyed them. Fleming described his ivory-bill in the following way:

> One of these birds was shot by my servant, which I took to be the hen, the feathers on the throat and belly and part of the wing and tail a shining black, it had nine stiff & strong feathers or pinions in the tail forked at the end, the middle one being six inches long from where the feathers begin the whole length being 7 1/2 inches the others on each side shortened in length, its wings ten inches long from the shoulder to the tip 18 long feathers in the wing, the two first and longest black the 3rd tip'd with white and

each succeeding one more till those next to the back are all white, both above & below, the front & fore part of the Crown black, from the junction of the upper & lower bills white feathers on each side, leaving a triangle of black feathers from the Eyes and back part of the Crown which is a deep red, the white feathers run backwards as far as the white on the wings intermixed with black so that the bird from the head so far appears speckled, the red part of the crown appears triangular, its legs was an inch & half long, with four toes set forwards & back two each way, armed with strong crooked claws, the two outer ones the longest & 4 inches in length [spread?] the bill white and bony, verry strong & firm at the point shaped like a wedge each $\frac{1}{8}$ of an inch broad and from that a ridge runs both in the upper and lower so that each forms a triangle an inch & quarter broad at the junction of the upper and lower bills, which is three inches in length, the tounge is six inches in length. The iris when dead of a bright Yellow so far it differs from any of the species I have seen, the mechanism of its parts being as usual in birds of this kind, it weighed upwards of 1 lb.[9]

William Bartram

Half a century after Catesby, William Bartram, a contemporary of Colonel Fleming, became the first American to devote his life to the study of nature. The son of colonial naturalist John Bartram, William was both a keen recorder of nature and one of the principal sources on Indian culture in eastern North America. However, Bartram provided little new information about the ivory-bill, noting that it was a resident species native to the Carolinas and Florida where he traveled extensively.[10] He described the ivory-bill as "*Picus principalis*; the greatest crested woodpecker, having a white back." The scientific name is that given by the Swedish botanist Carl von Linné—better known as Linnaeus. Linnaeus had described the ivory-bill for science based on the work of Catesby. Bartram's English description of the ivory-bill, however, was new and, in 1875, the eminent ornithologist Elliott Coues apparently thought it included a typographical error, suggesting that Bartram really meant "having a white beak."[11] It seems to me that Bartram wrote what he intended to write. The "white back" is an effect created when the ivory-bill folds the white secondary feathers and inner primaries of its wings over its back, and is the characteristic by which an ivory-bill can most easily be distinguished from a pileated woodpecker in the field.

Although providing little on the actual woodpecker, Bartram provided vivid descriptions of habitats he passed through as he traveled through the Carolinas, Georgia, and Florida. These descriptions suggest the nature of the world in which ivory-bills lived before the clearing of the virgin forests. For example, in the spring of 1773, near Wrightsborough, Georgia (now the town of Thomson, just west of Augusta), Bartram wrote of

> a perfectly level green plain, thinly planted by nature with the most stately forest trees.... many of the black oaks [*Quercus velutina*] measured eight, nine, ten, and eleven feet diameter five feet above the ground, as we measured several that were above thirty feet girt, and from hence they ascend perfectly straight, with a gradual taper, forty or fifty feet to the limbs.[12]

John Abbot

In 1751, two years after Catesby's death, John Abbot was born in London, the son of an attorney.[13] As a youngster he was fond of catching butterflies, and he busied himself with sketching and painting them. His parents were interested in art and nature and had a large collection of fine art and a good library. As young Abbot's interests in natural history and art developed, his father arranged for drawing lessons. His teacher was also interested in insects, and one good mentor led to another, guiding Abbot's growth as a student and artist of nature. Through his own purchases and gifts from his father and others, Abbot's personal library grew to include the most important works on insects and birds, including books by Catesby.

As he matured, Abbot studied law but continued his natural history art, earning good fees by painting rare insects for collectors. In 1773, at the age of 22, Abbot decided to travel to New Orleans so that he could paint American insects and birds. But after reading of Virginia, he sold his collections, earned additional funds by painting shells for a collector, and in July left for Virginia.

On about September 9 he arrived near Jamestown, Virginia, where he stayed for about two years. It was three years before the Declaration of Independence and political problems between England and the American colonies were coming to a head. Abbot, now in a hotbed of insurrection, had no intention of fighting for the rebelling colonies; he left Virginia and traveled south to the neutral territory of Georgia. At first he lived on a plantation about thirty miles south of Augusta, but eventually he moved farther south to Savannah.[14] Abbot's early Georgia years, just south of Augusta, likely gave him his first experience with ivory-billed woodpeckers. Burke and Screven Counties in Georgia border the Savannah River and the nearby Savannah River Swamp that likely harbored ivory-bills into the early twentieth century.[15] As the political situation in the country degraded to war, Abbot finally aligned himself with the colonists and fought in the American Revolution. For his service he was given a land grant of 575 acres on the Ogeechee River in Georgia. He resided in Georgia for the remainder of his life. He lived where he studied—it was a daily pursuit, unencumbered by a need to travel about the countryside selling subscriptions as Wilson and Audubon later did. In addition to other natural history subjects, Abbot produced more than 700 bird paintings.

Abbot seemed to live the life of a country gentleman, studying natural history, collecting specimens, and drawing and painting them. We do not know for sure, but it is likely that his efforts were not solely to satisfy his own curiosity. He sold many specimens (mostly insects) and drawings to wealthy patrons in Georgia and Europe, so perhaps his bird art contributed to his income in a significant way.

By living in the same area for decades, Abbot had the opportunity to develop considerable knowledge and familiarity with local natural history and gain a reputation as the local expert. When Wilson came through the area selling subscriptions to his *American Ornithology,* he paid Abbot a visit and stayed for at least several days. Wilson wrote to

Figures 5.2 and 5.3. John Abbot's paintings of male and female ivory-billed wood-peckers. Date unknown; location likely Georgia. *Courtesy Houghton Library, Harvard University*

Bartram on March 5, 1809: "There is a Mr. Abbot here, who has resided in Georgia thirty-three years, drawing insects and birds. I have been on several excursions with him. He is a very good observer and paints well."[16]

Wilson recognized the quality of Abbot's work and Abbot gener-ously gave Wilson specimens and a wealth of information about south-ern birds. Abbot later corresponded with Wilson on a regular basis, sending him both information and specimens. Of interest is Wilson's response of January 23, 1812, to a letter he received from Abbot. In his letter Wilson commented on Abbot's report of a "Green-billed Woodpecker": "I do not know the large Green-billed Woodpecker—nor any woodpecker as large as the Wood Cocks—if you know of such, be so good as send me one."[17] Wood cock was a local name applied to pileated woodpeckers, thus one might speculate as to whether this "green-billed woodpecker" was an aberrant ivory-bill.

In the 1890s, a numbered series of Abbot's watercolor drawings was dis-covered in the library of the Boston Society of Natural History.[18] Of the original 200, 19 were missing. Two of those found were of a male (Figure 5.2) and a female (Figure 5.3) ivory-bill. This collection of Abbot drawings is now in the Houghton Library at Harvard University. Following the style of his British bird-artist friend, George Edwards, the birds are each portrayed in a flat profile on a stub, the base of which is adorned with unidentifiable foliage. Each portrait looks a bit like a stuffed bird fastened to a bonsai. The plates were labeled "White billed Woodpecker" (following Wilson). Watermarks on the paper suggest they were drawn before about 1810.[19]

Another set of 297 Abbot bird drawings completed in about 1815 has been preserved at Knowsley Hall, the British home of Lord Edward Smith Stanley, who purchased the drawings from a Georgia physician in 1817. An ivory-billed woodpecker is also included in that set and with it Abbot's brief comments on the bird: "This Species lives entirely on Insects, and chiefly in Swamps, barking the dead trees in a very dextrious manner in search of the larva of Beetles. Eggs 4 or 5 pure white, as large as pullets, equally thick at both ends."[20]

Abbot was not a prolific writer, and unlike Wilson and Audubon, there is no organized, published collection and analysis of his work. I think that Abbot might well have shared what he knew about ivory-bills with Wilson. Abbot, for example, was in a position to truly know that ivory-billed woodpeckers were nonmigratory. Wilson was not.

Louis Jean Pierre Vieillot

While Abbot was developing into a naturalist in England, Louis Jean Pierre Vieillot was undergoing a similar transformation in France. Born in Yvetot, France, on May 10, 1748, Vieillot's early employment was as a clerk in Paris, but his spare time was spent pursuing his avocation in ornithology.[21] When his family left France for a French colony on the West Indian island of Santo Domingo, Vieillot went with them, working as a businessman but continuing to collect birds. When the Napoleonic wars began, Vieillot, like Audubon later, was vulnerable to being drafted into the French Army. Apparently to escape the draft he moved to North America from the French colony and may have remained for several years. About 1807, Vieillot published his two-volume work, *Histoire Naturelle des oiseaux de*

Figure 5.4. The eggs of ivory-billed *(top left)*, pileated *(right)*, and red-cockaded woodpeckers *(bottom)*. These eggs are all from the ornithological collections of the Western Foundation of Vertebrate Zoology (WFVZ #6856-2; collected February 16, 1905, by R. D. Hoyt and John R. Davey).

l'Amérique septentrionale, which included 131 plates of the birds of North America.

The ivory-billed woodpecker was included in volume 2 of Vieillot's *Histoire Naturelle des Oiseaux de l'Amerique Septentrionale.* His plate of the ivory-bill (Plate 1) shows a male with a greatly exaggerated crest standing atop a stump. Vieillot also provided a detailed drawing of the bill of an ivory-billed woodpecker shown in comparison with the bills of other birds in volume 1 of the same work. The great tragedy of Vieillot's work is that there are no known journals of his travels, and his writings only infrequently indicate localities. Thus we have no idea where he saw (and probably collected) ivory-billed woodpeckers.

Alexander Wilson

Alexander Wilson, often referred to as the "father of American ornithology," was born in Paisley, Scotland, in 1766.[22] A self-taught painter and the son of a Scottish smuggler, Wilson spent his youth as a weaver, a peddler, and a poet. He also championed the working class, but his efforts led to his arrest for blackmailing a cloth buyer in Paisley who was using a yardstick longer than a yard to cheat the weavers. While out on bail after a second arrest, he fled to North America, arriving in Delaware on July 14, 1794. Wilson quickly gained employment in Philadelphia as an engraver, a trade with which he had no experience. Within a few months, he turned again to weaving and in the spring of 1795, he peddled his cloth through northern New Jersey. He then took a position as a teacher.

Living near the Delaware River, Wilson also became enthralled with the annual migrations of waterfowl and the abundance of songbirds in nearby woods. He kept a wide range of native birds as pets. In 1801, Wilson's world closed in on him. He had apparently had an affair with a young lady and something happened such that he felt he had to leave town in a hurry. He did so, leaving his belongings behind and contacting friends without revealing his whereabouts. Distraught over his personal life, he longed to return to Scotland. Wilson's biographer Robert Cantwell referred to him at this time as being a "spiritual convalescent"—it seems that everything he tried ended in failure.[23]

In February 1802, Wilson accepted another teaching position, at Gray's Ferry, across the Schuylkill River from Philadelphia, but he became reclusive, afraid to allow close friendships lest they end as before. Instead, he read and wandered the countryside and studied the natural world around him. Among his regular haunts was the wooded garden of the colonial naturalist William Bartram. Although at first he visited only the garden, he later became friends with Bartram and seems to have used Bartram as an artistic mentor.

In 1803 Wilson began an extraordinary effort. On June 1, 1803, he wrote to his friend, Thomas Crichton in Paisley, Scotland, that he had a new goal: he was going to make a collection of "all our finest birds."[24] Wilson was a man with a plan, and he pursued that plan with incredible vigor. His interest in birds became known throughout Philadelphia, and Wilson came under the influence of Charles Willson Peale, an artist and founder of Peale's Museum. Both Peale and Bartram encouraged Wilson in his efforts.

Through the 1803 school year, Wilson's students brought him birds, mice, opossums, and other creatures to draw. Wilson worked feverishly on his project, cutting costs by doing his own engraving of plates for mass production of prints. In April 1806 he resigned from his teaching position to accept a better-paying position as assistant editor for the twenty-two-volume *Ree's Cyclopaedia.* The editorial experience would contribute to his abilities to complete his own book project. At the same time publisher Samuel Bradford agreed to publish the volumes of Wilson's *American Ornithology.*

As Wilson prepared his sample plates he also began using his skills as a peddler, traveling the countryside studying, sketching, and painting new birds while selling subscriptions to the anticipated ten-volume *American Ornithology.* When Thomas Jefferson purchased a subscription, doors opened and subscriptions picked up.

Everything that Wilson had done during his life came to bear on his success in producing *American Ornithology.* His experience as a teacher, a peddler, a poet, a self-taught artist, and an engraver all contributed to his final masterpiece. His goals had been to discover new birds for inclusion in his work and to sell subscriptions. The latter seems to me to have been the greater goal, but in spite of his obvious desire to sell his work and a lack of formal educational training, Wilson was a scholar. He learned from personal observations, but he also was skilled at obtaining and making use of information gleaned from others. Everywhere he went he inquired about birds. When he found a knowledgeable person, he took every opportunity to learn as much from him or her as he could, fully acknowledging this person's contributions to his scholarship.

During the winter of 1808 to 1809, Wilson headed south, leaving Washington on Christmas morning. In early February, while crossing what is now called Holly Shelter Swamp twelve miles north of Wilmington, North Carolina, Wilson first saw ivory-billed woodpeckers. He had read Catesby's account of the species written a century earlier, but could hardly believe that a woodpecker might produce a bushel of wood chips at the base of a tree in an hour. Now he believed. The bird was magnificent. He found not one but three ivory-bills and shot them all, killing two and wounding the third. He was ecstatic to have secured a living male. He described this first encounter and his ensuing experience with the wounded bird in great detail in *American Ornithology,* providing an engaging narrative that has been repeated by many authors.[25]

Wilson wrapped the bird in his coat, mounted his horse, and continued into Wilmington. En route to Wilmington, the bird continually gave distress calls—so loud and shrill that they spooked his horse into running off into the swamp, nearly throwing Wilson and his captive. As he rode into town, the ivory-bill continued crying, likened by Wilson to the piercing scream of a small child. People came to their doors and stopped to peer at the strange man and his charge.

As he dismounted at an inn, the bird cried out again and the innkeeper and guests rushed out to see the problem. Wilson tells of how he asked for accommodations for himself and his baby and at first received rather quizzical expressions in response. Then he opened his coat to reveal the bird and the group had a good laugh. Wilson left the ivory-bill in his room while he tended to his horse. He was gone less than half an hour, but when he returned, his bed was covered with plaster and there was a hole the size of his fist nearly through the wall adjacent to the window. He recaptured the bird, tied a string to one leg, and tied the other end of the string to a table leg. He wanted to keep the bird alive and left the room in search of food for it. On his return he discovered that the woodpecker had nearly destroyed the mahogany table.

At that point Wilson decided to paint the bird before something else happened and it escaped or died. As he worked with the bird, he was struck several times and severely cut. I can well understand the pain, having incurred the wrath of several species of woodpeckers that I have held in my hand during banding. Even the small downy woodpecker can quickly draw blood. And when a woodpecker is striking at an adversary, its bill is usually slightly opened, thus providing two sharpened weapons and creating two puncture wounds with each stab. Wilson said that he painted the ivory-bill in several poses, but his Plate 29 includes only three profiles plus profiles of pileated and red-headed woodpeckers for size, shape, and plumage comparison. He noted that the locals knew little about the ivory-bill and did not distinguish it from the pileated except to call the ivory-bill the "Greater Logcock" and the pileated the "Lesser Logcock." He also noted that they rarely shot one. Wilson's ivory-bill died about three days after he began his paintings.

On the same trip Wilson also encountered ivory-bills in a cypress swamp bordered by longleaf pines (*Pinus palustris*) across the Savannah River in South Carolina near Augusta, Georgia. I have spent a great deal of time in the Savannah River Swamp and have seen several extremely old baldcypresses of enormous size—likely trees that were present when Wilson was there. The swamp is still extensive in size and serves as a buffer for what is today the Savannah River Site, a nuclear facility where plutonium is processed for weapons. Unfortunately (for ivory-bills) most of the forest is young and the adjacent pinelands are mostly young pine plantations.

Wilson almost certainly encountered the ivory-bill in other localities during his travels. With copies of his first volumes to show, in the spring of 1810 he embarked on another trip to sell subscriptions and collect new birds and drawings. He headed southwest through Pennsylvania and Ohio, then south through Kentucky and Tennessee, taking the Natchez Trace from Nashville and through Mississippi to New Orleans. Wilson's return trip was by ship to New York. Shortly after his return from New Orleans he responded to a letter from John L. Gardiner apparently telling him of a possible new woodpecker he had found. Gardiner's woodpecker was apparently a pileated. In his response on September 12, 1810, Wilson noted, "The Woodpecker you mention I am already acquainted with. It is very numerous in the Southern states, and even in the Gennesee Country, where it is called the Woodcock. There is a still larger species found in Louisiana, called the ivory billed Woodpecker, the biggest of the whole genus, both of these will appear in Vol. IV."[26]

As he promised Gardiner, Wilson's illustrations of the ivory-billed woodpecker were included in Volume 4 of *American Ornithology* in Plate 29 (Figure 5.5). His contributions to the knowledge of ivory-billed woodpeckers are significant in two ways. First, it seems he is the only person to have kept a live "adult" ivory-bill in captivity for any period of time—although his bird was wounded and lived only a few days. Nonetheless, Wilson was in a unique position. He had kept many species of birds in captivity for varying periods of time, thus he had some familiarity with aviculture. He also knew that he could learn from this captive, and he seems to have had the intuitive capacity to maximize what he could learn from his observations. I have often kept and cared for injured birds and the insights I have gained have led me to tell students that they will never truly understand a bird until they have lived with it. Thus I give great credence to Wilson's comments on distress calls, climbing behavior, the way it held its body, the strength of its feet, and details in his illustration of ivory-bills.

The adult male ivory-bill at the upper right in Wilson's plate illustrates a characteristic that has not been illustrated by any other artist. There is a white spot shown in the crest just behind and above the eye. In *American Ornithology*, Wilson described the plumage of the male as including the "fore part of the head black; rest of the crest, a most splendid red, spotted at the bottom with white, which is only seen when the crest is erected." None of the other colonial naturalists, and no modern ornithologist, has mentioned this white in the male's crest, although Walter Hoxie, in 1887, quoted a native hunter in coastal South Carolina as mentioning having seen "one of the old-fashioned big woodpeckers with a red and white top knot."[27]

I had never noticed such a spot and looked for it the next time I exam-

Figure 5.5. Plate 29 from volume 4 of Alexander Wilson's *American Ornithology* (Philadelphia: Bradford and Inskeep, 1808–1814). Figured are ivory-billed, pileated, and red-headed woodpeckers. *Courtesy Albert M. Greenfield Digital Imaging Center for Collections, Academy of Natural Sciences, Philadelphia*

ined ivory-bill specimens. I was surprised to find that it does exist (Figure 5.6), and only in males. The base of each red crest feather is strikingly white. My guess is that the white might be shown during aggressive encounters when two birds are squared off at one another. In other woodpeckers the feathers of the head are fully erected during such encounters, thus maximizing the apparent size of the head. Such raising of the crest is also characteristic of a captured bird that is held in the hand and protesting against its captivity—which is probably how Wilson saw the display.

Other contributions from Wilson's writings include his measurement of wingspan at thirty inches; his description of eye color as vivid yellow, of the tongue as "worm-shaped, and for half an inch at the tip as hard as horn, flat, pointed, of the same white color as the bill, and thickly barbed on each side"; of leg color as "light blue or lead color"; of the stomach as "an oblong pouch" not like the gizzard of others; and of calls similar to "the tone of a trumpet, or the high note of a clarionet."[28] Within Wilson's account we also find information that he simply could not have known from his own experience—such as the species range and nonmigratory status. As noted earlier, the collective knowledge of Abbot and others likely contributed to Wilson's understanding of such things.

Over the course of his life, particularly the final decade, Wilson had described, drawn, and painted 264 species of North American birds. In August 1813 he died of dysentery. He had completed eight of the ten volumes he had proposed, with the ninth published a year after his death. Wilson succeeded in his last great endeavor, setting high standards for the science of American ornithology.

John James Audubon

About the same time Wilson was coming of age, Audubon was entering the world. There is a certain mystery that hangs over the birth of Audubon, perpetuated in part by him, by his heirs, and by a lack of records. An unlikely yet intriguing version, written by a distant relative, holds that he was the son of Marie Antoinette and Louis XVI.[29] Most biographers, however, suggest that Audubon was born on April 26, 1785,

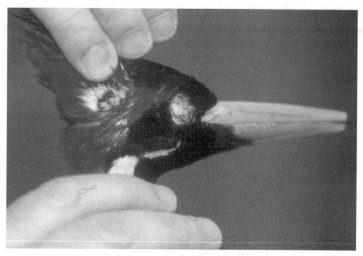

Figure 5.6. The head of a male ivory-billed woodpecker specimen with the red crest feathers parted to show their white base as portrayed by Alexander Wilson. Specimen at the US National Museum. *Courtesy J. A. Jackson Jr.*

at Les Cayes, Santo Domingo, the son of a French sea captain and his Creole mistress, Mademoiselle Rabin. Audubon's mother was killed during a slave uprising shortly after his birth, and young John was taken back to France where Captain Audubon's wife adopted him.

Audubon loved the outdoors as a child, and it has been suggested that he studied art before his father sent him to North America to manage his property and to avoid Napoleon's draft. Audubon's father had visited North America on many occasions and had served with the French troops who supported the American Revolution. He had purchased the Mill Grove estate along Perkiomen Creek in Pennsylvania, where Audubon's serious interest in birds probably began. But instead of turning to natural history as a vocation, it remained an avocation while Audubon demonstrated his ineptitude as a businessman. When he was operating a failing store in Louisville, Kentucky, Audubon was visited by Wilson, who stopped by to try to sell him a subscription to his *American Ornithology*. Audubon was on the verge of subscribing when his partner dissuaded him, suggesting that Audubon's art was much better than Wilson's.

With the failure of his business and a growing family, he had a new goal perhaps stimulated by Wilson's visit: to paint all the birds of North America. Audubon seemed to succeed only as an artist, and thus set off on his journey to success and fame. Unable to find a publisher and engraver in North America (they had all invested in Wilson's efforts), Audubon ended up working with Robert Havell in London to produce his *Birds of America*. His travels took him through much of North America—much farther and with much more time spent afield, especially in the range of the ivory-bill, than Wilson had gone.

Audubon carefully recorded his observations and produced drawings that were much more lifelike than any before him. Like Wilson, he had to sell subscriptions to his work. And like Wilson, his travels served double duty—getting subscriptions and new material for his books. But unlike Wilson, Audubon did not work alone. His sons and a number of others traveled with him at various times and contributed both to his drawings and to his understanding of birds.

Ivory-billed woodpeckers are mentioned several times in Audubon's journals of 1820 to 1821. On October 12, 1820, Audubon left his wife and family in Cincinnati and, accompanied by an apprentice, thirteen-year-old Joseph Mason, paid for passage on a flatboat bound for New Orleans. He planned to be gone for seven months, collecting and painting birds. On November 17 they reached the Mississippi. Two days later they passed Wolf Island and Audubon noted, "Ivory Billed Wood Peckers are Now Plenty."[30] On November 21 they landed at New Madrid, Missouri, and Audubon noted that the woods "have Nothing in More than the *Pait Pait Pait* of the Monogamous—Wood Peckers—" apparently a reference to ivory-bills, because he later described their call note as *Pait* and often remarked on how they traveled in pairs.[31] On November 24, he again mentioned ivory-bills, noting "these birds allways go in Paires and when they Leave a Tree to fly to another they Sail and Look Not unlike a *Raven*."[32]

On December 14, below the mouth of the Arkansas River, Audubon saw five ivory-bills feeding on berries of some creeper and he noted they were "gentle—Keeping a Constant Cry of *Pet Pet Pet*."[33] On December 17, near Pointe Chico in what is today southeast Arkansas, he noted that they were becoming "more plenty." On December 20, Captain Sam Cummings, a member of his party, shot an ivory-bill but only broke its wing. When they tried to retrieve it, the bird hitched to the top of a tree and they thought it had escaped until Mason brought it down with a single shot. The next day they reached Stack Island in northeast Louisiana, and Audubon wrote of the "constant Cry of ivory Billed Wood Peckers about us—scarcely any other except a few *Peleated* [= pileated] and Golden Wings [= yellow-shafted flickers (*Colaptes auratus*)]." [34]

The following summer, on August 12, near Bayou Sarah, Audubon saw several pairs of ivory-bills and killed a male. That day he had described traveling through tall baldcypress and rich southern magnolia woods.[35]

Audubon's account of the ivory-bill in the text of the octavo edition of *The Birds of America* provides a thorough view of the behavior, ecology, and anatomy of the species.[36] Audubon's work and that of James Tanner in the 1930s provide us with most of what is known about the lives of ivory-billed woodpeckers. The magnitude of Audubon's contribution to our understanding of the species can be appreciated by realizing that Audubon lived much of his life within the range of ivory-bills. His was not a brief encounter but a life with ivory-bills.

Audubon provides the clearest picture of the ivory-bill's flight and many other aspects of its behavior, as well as its distribution in Colonial America and its anatomy. The portrait of the ivory-bill from Audubon's *Double Elephant Folio* (Figure 5.7) includes a male and two females in life-like poses, not merely of proper proportions and color pattern but also showing them as he knew them—social, traveling and foraging in family groups, feeding on large beetles and their larvae.[37] Audubon also provided an anatomical view of the head and neck of the ivory-bill (Figure 5.8).[38] This and other anatomical drawings were completed by William MacGillivray, a young Scottish university professor of comparative anatomy, who also contributed greatly to the text accompanying *The Birds*

Figure 5.7. Painting of ivory-billed woodpeckers by John James
Audubon. Done before 1826 and labeled "Louisianna," this
plate that includes a male and two females appeared in Robert
Havell's edition of Audubon's *Birds of America* (known as the
Double Elephant Folio) as plate number 66. *Courtesy Albert M.
Greenfiled Digital Imaging Center for Collections, Academy of
Natural Sciences, Philadelphia*

of America by correcting Audubon's deficiencies in English.[39] There is no
species that Audubon illustrated more thoroughly than the ivory-bill.
Aware of the work of his predecessors, Audubon learned from them, built
on their efforts, and went far beyond what they knew. With the possible
exception of Tanner, no other individual has given us so much of what we

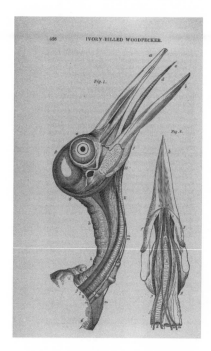

Figure 5.8. Anatomical drawing of an ivory-billed woodpecker apparently done by William MacGillivray for John James Audubon's *The Birds of America. Courtesy Albert M. Greenfield Digital Imaging Center for Collections, Academy of Natural Sciences, Philadelphia*

know about the ivory-billed woodpecker. Indeed, much of what I have written could not have been conceived had Audubon not failed in business and found profit in the study of birds.

Colonial America was a small world both in terms of area occupied by the colonists and in terms of human populations. The links among influential people of the eighteenth century are often amazing. Audubon and Wilson, for example, not only met at Audubon's store in Louisville, Kentucky, but also shared many friends and acquaintances in Philadelphia and elsewhere. Both knew and were influenced by Abbot, and each gained much of his knowledge of New World natural history from Catesby and John and William Bartram. Together they have left a written and illustrated record of the world of the ivory-bill that we otherwise could only imagine.

CHAPTER 6

Arthur Allen

*I have just enjoyed one of the greatest experiences of my life, for I
have found that which they said could not be found—the Ivory-Billed
Woodpecker.*

—Arthur Augustus Allen, 1924

$\mathscr{I}$ know I had a teddy bear that was long-legged, khaki brown, with
lots of patches from hard wear. Special childhood toys often stick in
one's memory. Arthur August Allen had an unusual toy that he carried
around as a toddler: a stuffed scarlet tanager (*Piranga olivacea*).

Allen was born in 1885 and grew up in Buffalo, New York, where he
was an accomplished naturalist even before he entered Cornell University
as an undergraduate in 1904.[1] He had the benefit of a father who was both
an attorney and an amateur naturalist. His career might have taken a turn
toward fish as a result of his work as an assistant to the famed ichthyol-
ogist David Starr Jordan or to mammals after he completed his master's
thesis at Cornell on the mammals of the Cayuga Lake Basin. But it did
not. He remained at Cornell to complete a doctoral thesis in 1911 on red-
winged blackbirds (*Agelaius phoeniceus*). In 1911, after receiving his Ph.D.,

Allen spent a year in Colombia collecting specimens for the American Museum of Natural History. The assignment completed, he returned to Cornell to teach ornithology. In 1915, Cornell created the first North American graduate program in ornithology and promoted it under the name "Laboratory of Ornithology," with Allen as its head. It was a conceptual laboratory; there would be no physical laboratory building until 1957. But Allen was spreading the word. By 1943, he had taught ornithology to more than 10,000 students.[2]

Ivory-Bills in Florida

Allen is well-known for many ornithological reasons. In addition to being the founder of the Cornell Laboratory of Ornithology, he wrote extensively for popular audiences and was responsible for initiating both undergraduate and graduate curricula in ornithology. But Allen was also a principle figure in seeking out and documenting the survival of the ivory-billed woodpecker in the early decades of the twentieth century.

In the winter of 1924, Allen and his wife Elsa spent a university sabbatical seeking rare birds in Texas and Florida. One of the birds he hoped to find was the ivory-billed woodpecker. W. H. Mann, of St. Cloud in central Florida, had seen four ivory-bills along Taylor Creek in Osceola County.[3] Allen had held little hope of finding the ivory-bill, nor did the most knowledgeable of Florida birders feel that it could be found. But, in April of 1924, after a month of searching, Allen proudly announced its discovery in *Bird-Lore* magazine.[4] He and his wife and their guide, Morgan P. Tindall, found a pair of ivory-billed woodpeckers near Taylor Creek in northern Osceola County in central Florida. Allen photographed the pair on a dead stub (Figure 6.1). On a copy of the photo in the archives at Cornell University Allen wrote on the back: "The Ivory-billed Woodpecker will soon follow the *Archaeopteryx* to extinction."[5]

The celebrated discovery soon turned bittersweet. Allen needed to leave Florida but planned to return and spend most of the next month studying the birds; however, on the day he left for home, he learned that the son of a man who had been putting a fence through the bald-

Figure 6.1. Ivory-billed woodpeckers photographed in Florida by Arthur A. Allen on April 13, 1924, in central Florida. This photo is the first photo ever taken of a living ivory-billed woodpecker and is the only photo of an ivory-billed woodpecker at its nest taken in the United States except for Tanner's and Allen's later photos from Louisiana. Shortly after the photo was taken, local taxidermists collected these birds. They were ultimately sold to the Florida Museum of Natural History. *Courtesy Arthur A. Allen II*

cypress swamp had led two taxidermists into the area. They had shot two ivory-bills.

On June 19, 1924, Allen wrote to Tindall to try to get confirmation that the birds he had observed were the ones killed.[6] He told Tindall the taxidermists' names were Hancock and that they would not respond to his attempts to contact them. Apparently the birds had been collected under a permit from the state.[7] Tindall replied on June 29 that he had gone to see the fence builder concerning the killing of the ivory-bills but that the fence builder could not answer with certainty. The fence builder said that he supposed someone must have killed the pair of birds that Allen had been watching in the pasture, because they had not been seen again. He added that he had seen three pairs of ivory-bills at one time when he was building the fence through Taylor Creek Swamp, and suggested that by the next spring he could locate their nest.

The July 6, 1924, issue of the *Observer-Dispatch* newspaper of Utica, New York, carried news of the birds' discovery and loss under the headline "Cornell Man Finds Birds of Rare Kind," with the sensational subheading: "Gang of Taxidermists End Its Observation of Ivory-Bills by Shooting Entire Flock—Gets Movies."

I found additional details of the birds having been collected, but no evidence of a permit being issued. The only mention of a specific locality for where Allen found the birds is "Taylor Creek" given in correspondence between Allen and Tindall and in James Tanner's Audubon report.[8] Tanner told me that this was in northern Osceola County, but at the time he provided no county name in his report, and the information given suggests that this may have been the same locality at which a specimen in the Field Museum in Chicago was collected in 1907. On the specimen label Taylor Creek is listed as being in Orange County, but the creek also flows into Osceola County for several miles.[9] The previous entry in Tanner's report refers to ivory-bill records in 1916 on "Taylor Creek, Osceola County."

Examining specimens and their related correspondence at the Florida Museum of Natural History in Gainesville, I came across a series of letters between the museum and a Mr. B. G. Hancock of Loughman, Florida. These letters related to two ivory-bill specimens the museum purchased from him in 1928.[10] The museum paid $175 for the two birds, but there was considerable wrangling over the payment, and Hancock did not receive anything until about 1935.

Hancock said he had collected the birds in "about June 1925" in a cypress swamp on "Bull Creek in Osceola County." The date of "about June 1925" was not with the specimens when they were acquired by the museum but rather provided by Hancock in a letter dated July 30, 1932. Tanner mentioned that the specimen associated with Allen's ill-fated 1924 ivory-bill venture was at the University of Florida Museum (now the Florida Natural History Museum), although there are no specimens or catalog records indicating the presence of Taylor Creek specimens at the museum.[11] I suspect that the Hancock specimens are the birds that Allen had found and that the correct date and location of collection should be "Taylor Creek, about June 1924."

A Team to Record Bird Sounds

Allen was a "public" ornithologist. He was not the ivory-tower type who spent his days working with specimens; nor was he the preeminent field person who ignored all else in favor of staying afield as much as possible to collect every last minutia of data. He was first and foremost an educator. His life's work was to educate the public about birds, their lives, and their needs. Although he made many original contributions to the understanding of birds, most of his work was anecdotal and focused on interpretation and documentation rather than on the analytical; his work was innovative rather than lengthy or detailed.

In 1929, Peter Paul Kellogg, a graduate student of Allen's, was appointed as an instructor in ornithology. This was fortuitous because Kellogg had a strong interest in sound-recording equipment. In May 1929, Fox-Case Movietone Corporation of Hollywood came to Allen for help in recording bird songs on movie film. Allen saw the possibilities for science and public education and Kellogg had the background to work out the electronics. They agreed to work with the studio.

In the fall of 1929, Albert R. Brand, a New York stockbroker who had quit his job to come to Cornell to pursue his avocation—bird study—joined the Allen and Kellogg team. In addition to his enthusiasm, Brand brought funds, making the development of efficient recording of bird songs possible.[12] Brand began buying and testing new equipment, and the number and quality of bird recordings increased dramatically. At the 1931 meeting of the American Ornithologists' Union (AOU), Brand presented a "Preliminary Report of a New Method of Recording Bird Song (Illustrated by Sound)." The presentation included recordings of thirty species of birds and was an incredible success. AOU secretary

T. S. Palmer referred to it as "the outstanding contribution" of the meeting.[13] Allen, Kellogg, and Brand were excited and began to explore possibilities for their new techniques.

Ivory-Bills in Louisiana

An article about ivory-bills by Georgia forester Charles Newton Elliott in the April 1932 issue of *American Forests* magazine must have given Allen a brief surge of adrenaline—and then resigned frustration.[14] Elliott claimed the presence of ivory-billed woodpeckers in the Okefenokee Swamp in south Georgia, but had presented only secondhand, although graphic, descriptions. The ornithological and conservation community cried "foul," or as Aldine R. Bird put it in 1932, they said "Show us or we cannot believe. . . . The King is dead."[15]

By the time of Elliott's publication, proof was already in hand—literally—that the rumors of the "King's" death had been premature. A discussion suggesting that the ivory-billed woodpecker was extinct had taken place in the offices of the Louisiana Wild Life and Fisheries Commission in New Orleans in the spring of 1932. Present at the time of the discussion was Mason D. Spencer, an attorney and state legislator from Tallulah in northeast Louisiana. Spencer insisted that the birds were not extinct and that they could still be found in the old-growth, floodplain forest between the Mississippi and Tensas Rivers in Madison Parish, just south of Tallulah.

But skepticism runs high when it comes to ivory-bills, and Spencer was not believed. Confident that there were no ivory-bills and that a pileated woodpecker would be brought in, they issued Spencer a permit to shoot an ivory-bill. After all, it had been in July 1899, thirty-three years earlier, that the last ivory-bills had been officially reported from Louisiana. George Beyer of Tulane University had then collected three ivory-bills just southwest of the Singer Tract.[16]

On about April 15, 1932, Spencer shot a male ivory-billed woodpecker in the general area that came to be known as the Singer Tract in Madison Parish, Louisiana, and delivered it to the Louisiana

Department of Conservation office.[17] Armand Daspit, director of the Fur and Wildlife Division of the Department of Conservation, immediately ordered that no more permits be issued for collecting ivory-bills and assigned a game warden to protect any that remained—once the word got out, collectors might be a problem. Taxidermist E. S. Hopkins prepared the bird for public display.

The National Association of Audubon Societies quickly responded. In less than a month, from May 12 to 17, 1932, Audubon president T. Gilbert Pearson and Ernest G. Holt, a Louisiana resident and Audubon's newly appointed director of sanctuaries, went to the Singer Tract to assess the situation. Assisted in their search by state officials, Pearson and Holt located at least one male and two females, and possibly as many as six individuals.[18] When word of the sightings reached Allen, his excitement grew. When his friend Pearson confirmed the sightings, Allen's excitement was uncontainable. As in 1924, Allen was once again eligible for a sabbatical, and this time he felt he would succeed in studying the ivory-bill.

In mid-February 1935, Allen returned to central Florida on the first leg of an expedition to record the voices of vanishing birds. This trip had been encouraged and sponsored by Brand and was a joint effort of Cornell and the American Museum of Natural History. Allen noted that the expedition also had the blessing of the National Association of Audubon Societies. The ivory-bill was only one of several troubled species the group sought, and they had planned the timing of each stop of their expedition to maximize their potential for success. Allen had felt that February was too early to find ivory-bills nesting in Louisiana but that they might be able to find them in central Florida. The field party included Brand and Allen as leaders, Kellogg as sound technician, George Miksch Sutton as artist and ornithologist, and then graduate student Tanner as general handyman to assist with both photography and sound recording.[19] A month of searching yielded no evidence of ivory-bills in Florida, and the party headed for the Tensas River swamp in northeast Louisiana.

Accompanied by Spencer and State Conservation Department warden J. J. Kuhn, the party located an active ivory-bill nest forty-three feet up in a dead swamp maple (red maple). The nest was seven miles from an improved road in the midst of an unbroken forest that extended eighteen miles from east to west and thirty miles from north to south. The mayor and sheriff of Tallulah volunteered use of their jail yard as a place they could dismantle the party's sound truck and install the sound equipment in a wagon that could be pulled into the swamp forest by mules. They had an audience for the effort—amused inmates watched, some volunteering that, if it could be "arranged" for them to get out, they could show the party more of these big "peckerwoods."

The sound wagon, pulled by four mules, was taken to within three hundred feet of the ivory-bill nest, and Allen and his party went to work (Figure 6.2). They borrowed the genus name of the ivory-bill, christening their camp almost literally in the shadow of the ivory-bill nest tree "Camp Ephilus." The party remained for eight days at the nest, with the birds becoming acclimated to the observers' presence and even tolerating a blind that was built in the top of an elm tree only twenty feet from the nest.

Last Chances and Fading Memories

On January 24, 1946, Allen wrote to his old friend Kuhn in Tallulah, Louisiana. He said he was planning another sound-recording expedition that would take them through Louisiana and added, "If you could guide us to a pair of birds as you did back in '35, we certainly would make it worth your while. . . . I should appreciate hearing from you very much as to whether it would be worth while to hunt for ivory-bills in your country, and it certainly would be a pleasure to see you again."[20] There was apparently no response; there seems to have been no sound-recording trip; and there may have been no ivory-bills.

Allen's experiences in the Louisiana forest recording the calls of ivory-bills and observing their behavior was recorded on motion-picture film. At the 1935 meeting of the American Ornithologists' Union, Kellogg presented a forty-five-minute film of their efforts with the ivory-bill. Would that we still had that film and Kellogg's narration

Figure 6.2. J.J. Kahn (Left) watches as Paul Kellogg (right) recods the sounds of ivory-billed woodpeckers at their nest in the Singer Tract in northeast Louisiana, April 1935. *Courtesy James T. Tanner*

of it. Unfortunately the thousands of feet of movie film shot by the Cornell team were on old nitrose film—film that was highly flammable and dangerous to keep around. Following the flurry of ivory-bill work in the 1930s, the film was put into storage at Cornell—apparently in an isolated shed or small building because of the danger it posed. Many years later, James Tate, Jr., then assistant director of the Cornell Laboratory of Ornithology, told me that when the film was rediscovered, it was deemed too dangerous to keep around, so a small sample of it was copied onto safer film and the rest was destroyed. Part of the footage was included in filmmaker Marty Stouffer's 1975 documentary *At the Crossroads—The Story of America's Endangered Species.*[21]

Allen twice "rediscovered" ivory-billed woodpeckers for science, and in eight days of intensive observation in Louisiana he provided a glimpse of their home life and recordings of their plaintive voice. In the end, Allen's ivory-bills haunt us as we watch and listen to them, setting out against all odds to produce perhaps their last generation.

CHAPTER 7

James Tanner

. . . an ability to rough it and to get along with all kinds of people in all kinds of situations, a natural adaptability, ingenuity, originality, and a willingness to work. Above all he had shown a clear mind and superior intelligence.

—Arthur Allen, describing why he had nominated James
 Tanner for the Audubon Fellowship to study the
 ivory-billed woodpecker

James Taylor Tanner was born on March 6, 1914, in Homer, in the heart of New York's Finger Lake region. As a small child his family moved to Cortland, about twenty-five miles northeast of Cornell University. At the age of 11, stimulated by his rural surroundings and an adult neighbor who encouraged him, he developed a keen interest in birds. As an undergraduate at Cornell, Tanner fell under the spell of Arthur Allen and his Laboratory of Ornithology. In October 1933, while walking near Ithaca, he discovered a black-capped chickadee (*Poecile atricapillus*) that had a totally black head instead of the usual "black cap." The result was his first professional publication, at the age of 20, "A Melanistic Black-Capped Chickadee."[1]

Introduction to Ivory-Bills

During the summer of 1934, Tanner worked as a naturalist at Glacier National Park in Montana, gaining self-confidence, field experience, and honing teaching skills. In December 1934, he finished his undergraduate work at Cornell. Because Cornell then had only spring graduation exercises, it was not until late spring 1935 that he graduated with honors in biology. In January 1935, however, he began graduate studies with Allen. In February 1935 he accompanied Allen, Peter Paul Kellogg, Albert R. Brand, and George Miksch Sutton on the Brand–Cornell University–American Museum of Natural History Expedition around North America to record the voices of vanishing birds.[2] That trip provided part of Tanner's master's thesis, *Sound Recordings for a Natural History Museum.*[3]

At the 1935 annual meeting of the American Ornithologists' Union at the Royal Ontario Museum in Toronto, Tanner presented a forty-minute film titled, *Some Birds of Florida and Louisiana,* a part of the Cornell lab's efforts to document rare species.[4] What an opportunity for an undergraduate—second only to the opportunity to have accompanied the Cornell team into the field! Tanner's film, however, immediately followed a forty-five-minute film presented by another member of the expedition, Kellogg, on *Recent Observations on the Ivory-Billed Woodpecker.* This was no easy act to follow.

Doctoral Studies

During the fall hawk migration of 1936, Tanner served as an Audubon warden at Cape May, New Jersey—an important link with the Audubon societies. On his return to the Cornell campus, he immediately began a doctoral program, hoping to study the behavioral ecology of the rare and elusive ivory-billed woodpecker. Allen encouraged the newly renamed National Audubon Society to provide support for the study and, after negotiations with Cornell University, Audubon established a graduate fellowship to support the research under Allen's direction. Allen then nominated Tanner for the fellowship.[5]

Tanner's dissertation work was to include searches for the ivory-bill throughout its North American range. Thus, on January 6, 1937, he headed south in his 1931 Model A Ford. His first destination was the Savannah River Swamp and bottomland hardwoods on Groton Plantation in Hampton County, South Carolina, just across the river and a bit south from Augusta, Georgia. While there, Tanner received a letter from Allen notifying him that the General Committee of the Cornell Graduate School had met on January 8 and approved his fellowship.[6] Allen forwarded the letter of appointment to him care of general delivery in Baxley, Georgia. The fellowship was for the amount of $1,500 per year. Tanner was to receive $300 for supplies, equipment, and any travel that Allen might have to make in association with the work, plus $100 per month stipend for the year of 1937. Ultimately, the fellowship was renewed for 1938 and 1939.

Although the sum of this fellowship seems minuscule by modern standards, it is worth noting that at the end of the three years, John Baker, executive director of the National Association of Audubon Societies, wrote to Allen that the unused funds would "certainly be helpful in connection with any publication of the ivory-billed Woodpecker research fellowship report."[7]

Tanner's frugality was matched only by Allen's stewardship of university funds. Before leaving for the South as his work began, Tanner had stopped by a local merchant and charged needed supplies to the university. In his letter of January 14, 1937, Allen noted that now that Tanner had the Audubon fellowship, he needed to send a check to him as reimbursement for those expenses.[8] The expenses included $5.36 for a knapsack, $1.05 for a "boy's axe," and 99 cents for a "lantern."

While Tanner was beginning his efforts in South Carolina, Baker was in the Singer Tract with state game warden, Jack Kuhn. Allen advised Tanner to visit old collecting localities in South Carolina, Georgia, Florida, Alabama, and Mississippi but to keep his focus on the Singer Tract. The Singer Tract seemed to hold the best hope for studying the birds and was a potential for a reserve built around the preser-

Figure 7.1. James Tanner as a graduate student in the Singer Tract. *Courtesy James T. Tanner*

vation of ivory-bills, even if hope was fading with every passing year. So Tanner headed for the Singer Tract.

In the course of his work with ivory-billed woodpeckers in the Singer Tract (Figure 7.1), Tanner also kept careful notes on other species. On May 3, 1937, he noted another of America's possibly vanished species, the Bachman's warbler.[9] He discovered a male "at the meeting of tall woods and cut-over land," and observed the bird singing, suggesting it was defending a breeding territory, until at least June 3, but saw no female. In 1938 he found no Bachman's warbler.

On December 24, 1937, Tanner conducted a Christmas Bird Count for the National Audubon Society in the Singer Tract, identifying birds in the field from 7 a.m. to 2 p.m.[10] In those seven hours he recorded 638

Figure 7.2. Nestling ivory-billed woodpecker, showing its dark eyes, short crest, short tail, and large feet. Photo by James Tanner. *Courtesy of James T. Tanner*

birds of 34 species, which is respectable for a single person on foot. Among the birds he tallied were two ivory-billed woodpeckers. The abundance of other woodpecker species he noted in the area may reflect Tanner's interest, but certainly reflects the quality of the habitat: seventy red-headed woodpeckers, nineteen red-bellied woodpeckers, nineteen pileated woodpeckers, eighteen yellow-shafted flickers, six yellow-bellied sapsuckers (*Sphyrapicus varius*), and four downy woodpeckers. Curiously, hairy woodpeckers did not show up in the count.

Two months later, in February 1938, Tanner found an ivory-bill nest in a dead stub of a living red maple and was able to climb to it by pounding spikes into the tree to use as steps. When he climbed the tree on February 24, he found a single nestling that he estimated was about two weeks old. He monitored the nest each day, and on March 6, decided that he wanted to band the bird so that he could identify it by sight after it left

Figure 7.3. Nestling ivory-billed woodpecker showing pattern of partially opened wing. *Courtesy James T. Tanner*

the nest. Having a banded bird of known age also offered the potential to learn how long ivory-bills might live, how far an individual might travel, and at what age it might first breed. With Kuhn to keep watch as he climbed, Tanner waited until both adults had flown out of sight and then climbed the fifty-five feet to the nest. The cavity entrance for an ivory-bill is so large that young can be retrieved by simply reaching in. As he grasped the young ivory-bill, the nestling struggled and squealed. Precariously perched, Tanner was able to slip the band onto the bird's leg and close it. He put the chick back into the cavity and then started to remove a small branch that had obscured his view of the nest entrance from the ground. As he was working on the branch, the nestling jumped from the cavity entrance and fluttered to the ground, landing in a tangle of vines where it grabbed hold and began squalling loudly.

Tanner hurried down the tree and grabbed the bird. Kuhn took it from him while Tanner reached for his camera. Worried that the adults would return, he started snapping away. The first six pictures were taken so fast that he had forgotten to focus. Calming a bit, he took several more carefully focused photos that clearly show nestling characteristics. Although one of the photos of the bird on Kuhn's head has been widely published, some, snapped in haste and less posed, show aspects of the bird and its plumage that have not been described (Figures 7.2 and 7.3), such as the pattern of the spread wing, the back with the wings open, and the abbreviated nature of the nestling's crest. With the film spent, Tanner wrapped the bird in two handkerchiefs, secured it carefully inside his shirt, and climbed to the nest. This time the bird remained in the cavity.

Tanner's efforts resulted in the only ivory-bill to ever have been banded. The Bird Banding Laboratory at the Patuxent Wildlife Research Center in Laurel, Maryland, identifies the nestling as wearing band number 365-27264, but no additional details are provided in banding files. A year later, Tanner saw the banded bird, still living within the territory of its parents.

Although Tanner only found ivory-bills in the Singer Tract, he spent part of each year searching for them elsewhere. In July 1938, Tanner visited Bessie Reid of Beaumont, Texas, to evaluate the validity of reports of ivory-billed woodpeckers in the Big Thicket area. He found that her reports of seeing ivory-bills were secondhand, then talked to the observers, but found their stories unconvincing. A few months later Tanner solidified his position as the ivory-bill expert when he gave a talk titled, "Food Habits of the Ivory-Billed Woodpecker" at the October 1938 meeting of the American Ornithologists' Union in Washington, DC.[11]

In 1939 Tanner was back at the Singer Tract, this time building a burlap blind forty feet up in a tree adjacent to the cavity in which he had banded the nestling. In his monograph on ivory-bills, he described what happened when he removed the blind, but his wife Nancy Tanner provided me with a more interesting anecdotal account:

> They flew and hopped from limb to limb of nearby trees, yapping and scolding. They repeatedly pounded and rapped hard on the solid limbs with their shining bills. In their short, excited flights they came to perches

within a few yards of mine. It was too much to have three adult Ivory-billed Woodpeckers shouting and knocking so nearby, glaring at me from their yellow-irised eyes, while I straddled the crotch of a tree some forty feet from the ground; *for the first and only time in my life there were too many Ivory-bills too near.* I covered myself again with the cloth of the blind.

On March 7, 1938, Tanner took photos in the Singer Tract—including the first color photo of a living ivory-bill (Plate 10). He took two other color photos of the bird, but I have been able to locate only one. According to Tanner's meticulous notes relative to the photos, this one is of an adult male at its nest that was thirty-five feet up. The photo was taken at 4:15 in the afternoon under a blue sky. The bird was photographed with an Agfa camera using Dufaycolor film shot at f8 at $^1/_{25}$th of a second. Dufaycolor was an early color film that was discontinued in the early 1940s. The color in the image today is quite remarkable, attesting to the film's durability. This image represents the best documentation we have of the pale yellow eye of the adult ivory-bill.

Tanner concluded his fieldwork in the Singer Tract on May 22, 1939. On June 15 to 16, he traveled to Georgia to check a report of an ivory-bill along the Canoochee River in Bryan County. After talking to the observer, he concluded that the bird was a pileated. He spent another day at the Savannah River Waterfowl Refuge in southwestern South Carolina near Savannah, Georgia. He then returned to his other great hope for the birds, the bottomland forest along the Santee River in South Carolina. The virgin forest of the Black Oak Island section of the bottoms seemed to be his best remaining hope to find ivory-bills, and he searched there from June 19 to 23 and again from November 29 to December 7. Although he found some possible signs of ivory-bills, the birds themselves eluded him. So having investigated forty-five areas and traveled 48,000 miles, Tanner returned to Cornell to finish writing his dissertation.

Continuing the Search and the Fight for Conservation

As he finished, ivory-bills remained a high priority. Tanner obtained a temporary appointment as a "collaborator" with the Civilian Conservation Corps under the National Park Service to make one more trip to the Singer Tract. He was there from March 23 to 29, 1940, to collect additional information on the current status of the area and data that might help with the possible establishment of a national park at the site. In the summer of 1940, Tanner published an article titled, "The Last Wilderness of the Mississippi Bottomlands" in *Louisiana Conservation Review,* complete with several photos showing the grandeur of the forest.[12] He spoke eloquently of sweetgums, southern red oaks (*Quercus falcata*), and honey locusts (*Gleditsia triacanthos*) five feet in diameter and of a cottonwood nineteen feet in circumference. He spoke of wolves, turkeys, bears, and panthers—it seemed he spoke of almost everything that might "sell" protection of the area. It is interesting to note, however, that he did not mention the ivory-billed woodpecker.

Following completion of his Ph.D., in the fall of 1940, Tanner accepted a position to teach biology at East Tennessee State Teachers College in Johnson City. There were only two new faculty hired at East Tennessee State that year; the other was Nancy Sheedy, a recent graduate of Harvard who had been hired to work in the psychology clinic where she taught diagnostic reading and psychology. The faculty at East Tennessee State always ate lunch together and Tanner and Sheedy struck up a friendship that blossomed into romance. They were married on August 15, 1941. On return from their honeymoon, Tanner had a draft notice waiting for him. Married men were not yet being drafted, thus he was reclassified. By the following May, however, the draft for World War II included married men, and the Army called him. A Navy commission arrived at the same time and Tanner opted to work in the Navy radar program. Tanner entered the Navy in mid-summer 1942 as a lieutenant, junior grade, and rose to the rank of lieutenant commander before being discharged in January 1946 to return to his teaching position at East Tennessee. Even as World War II raged on, Tanner continued his efforts on behalf of the ivory-bill, trying to garner support for protection of the Singer Tract as a park, a refuge—anything that would protect the remaining habitat.

In 1941, Baker asked to publish Tanner's dissertation as the first National Audubon Society Research Report. After considerable discussion and suggestions from Baker and others, "The Ivory-Billed Woodpecker" was published in the fall of 1942. The New York Zoological Society allowed Tanner to publish as a frontispiece their painting of an ivory-bill by Sutton. Roger Tory Peterson prepared some new maps for the report. Baker wrote a foreword praising Tanner's selection for the fellowship as a "happy" choice, and Allen wrote a preface commending the report as "the best working hypothesis we have for the possible preservation of this species that is balanced on the rim of extinction."

In December 1941, the National Audubon Society asked Tanner to return to the Singer Tract to check on the status of the ivory-bills. He was able to go for a two-week visit, this time accompanied by Nancy. She got her only view of ivory-bills this trip—a male and female as they returned to their roost holes. In a letter of May 20, 1992, Nancy described what she had seen:

A beautiful forest—virgin swamp timber with huge trees. The trees were bare so that we could see and hear quite far. The low places were covered with water, and I was wet most of the time!

My first view of an Ivory-bill was about dawn when we had arrived at the roost tree of an Ivory-bill that Tanner had already found. The trails were muddy and it was slow going in the dark, getting over logs, through vine tangles, wading through water, and slithering around mud-holes. But we got there in plenty of time to see a female come out of her roost hole and climb to the top of a stub and start rapping double raps and calling *kent.* She looked huge. Her shining black-and-white plumage, white bill, and bright yellow eyes made her very conspicuous. She was very noisy—her rapping and calling could be heard for a long distance. After a period of preening, she finally flew off rapidly, with Tanner in full pursuit. He called to me to stay where I was, but I wondered if and how he would ever find me again in that vast forest!

On January 4, 1942, Tanner wrote to Allen to report that because of the US defense program, logging near the Singer Tract had increased. It only affected the John's Bayou area, the heart of the Ivory-bill territory that had been cutover during summer and fall of 1941. He found an adult female and a juvenile female ivory-bill still using the area, but the male had not been seen since August. Tanner suggested that there had been no additional changes in ivory-bill numbers. Allen replied on January 7, lamenting that the logging operations had gone so far. With the war going on, Allen noted that there was little hope of saving the ivory-bills.[13]

Peterson and Bayard Christy, both prominent in ornithology and bird-watching circles, wanted to see ivory-bills before they were gone, and Tanner provided encouragement and guidance. More experts interested in the birds meant more voices that might help save them. On May 8 to 9, 1942, Peterson and Christy searched for ivory-bills in the Singer Tract.[14] Kuhn served as their guide as they criss-crossed the swamp, wading mostly along John's Bayou on May 8. Although they found extensively peeled bark on trees, the characteristic feeding sign of ivory-bills, they neither saw nor heard the birds. The next morning Peterson and Christy set out on their own, hiking and wading through the same area. Again they found peeled bark and knew there were ivory-bills in the area, but did not find them. At noon they were back along a logging road where they had seen feeding signs the previous day, when at last they heard an ivory-bill. Although it did not match the mental notes he had imagined based on reading the accounts of Audubon and Alexander Wilson, Peterson was certain of his identification. "What I heard was different, more of a *henk, henk*. An occasional blow would land—whop!—like the sound of an axe, as the woodpecker hammered a tree."[15]

At last they saw the bird, a female, half hidden by foliage as it swung its head right and left, testing with massive blows the soundness of the trunk on which it foraged. As it departed they had a clear view. Although ivory-bills can fly long distances, Peterson and Christy were able to follow the bird for awhile because each time it landed it called, revealing its location. Later they found a second female and followed it briefly. They thought they might have heard a third bird, but never saw more than two at one time. The birds fed low among the debris left from cutting in the area. They found no males, nor did either one see an ivory-bill again.

Putting the Ivory-Bill in Perspective

In 1947, Tanner took a position as assistant professor of zoology at the University of Tennessee in Knoxville and later rose through the ranks to professor. At the University of Tennessee he was instrumental in establishing a graduate program in ecology, but ivory-bills remained on his mind, and he often provided advice to those who would seek the bird. He continually evaluated reports that came in and often followed up on such reports on his own, with Nancy or in the company of other ornithologists. Tanner also frequently lectured to popular audiences about the ivory-bill and conservation of bottomland hardwood ecosystems, and he wrote for both popular and scientific audiences. In 1956, he published an article on the ivory-bill in *Texas Game and Fish*, stimulating considerable renewed interest in the birds in that region.[16] Tanner was great at putting things in perspective. For example, he noted in the Texas article that a tract of six square miles was needed to support one pair of ivory-bills, whereas that same tract also supported an estimated thirty-six pairs of pileated woodpeckers and 126 pairs of red-bellied woodpeckers.

Tanner, often accompanied by Nancy or their son David, canoed on many southern rivers, always with an eye and ear for ivory-bills. These trips included the Suwannee (1973), Wacissa (1975), Wakulla (1970 and 1975), and St. Mark's (1978) Rivers of Florida; the Altamaha (1965) and Satilla (1978) Rivers, and Okefenokee Swamp (several trips to Georgia between 1970 to 1989); the Congaree River of South Carolina (1980); and the St. Mary's River in both Georgia and Florida (1977 and 1980). None of their trips produced evidence of the continued existence of ivory-billed woodpeckers.

Tanner retired from the University of Tennessee in 1979 but remained an active and productive researcher in retirement, even reexamining his data from the Singer Tract.[17] He was always interested in learning more about ivory-bills. In 1986, the US Fish and Wildlife Service appointed Tanner, Lester Short of the American Museum, and myself to the Ivory-Billed Woodpecker Advisory Committee. On March 18, 1986, Tanner returned to the Singer Tract, invited as the guest of the US Fish and Wildlife Service as they were making plans for the headquarters and visitor center for Tensas River National Wildlife Refuge, which was at last going to protect what was left of the forest he loved. I am sure it was a bittersweet homecoming, saving what was left of the forest but doing so too late to save the ivory-bills.

Five years later, on January 21, 1991, Tanner died in Knoxville, Tennessee. I know from discussions with Tanner between 1986 and 1990 that he had come to believe that the ivory-bill is extinct in North America; nonetheless, he was always willing to listen and eager to share his expertise with those who held out hope.

CHAPTER 8

The Struggle for the Singer Tract

I believe that I have seen almost every bit of virgin timber in the South, and unreservedly, the Singer Tract has the finest stand of virgin swamp forest. . . . There is no question about it, the Singer Tract should be preserved.

—James Tanner, October 1939

By 1803, a few French and Spanish land grants had been issued for areas along the Tensas River in Madison Parish, in northeastern Louisiana, and some forest was probably cleared. After the Louisiana Purchase the number of settlers in the region increased, and as the first half of the nineteenth century wore on, rivers became the region's highways, and riverside forests were often cleared for fuel, cabins, settlements, and agriculture. Maps from 1846 show much of the course of the river lined with plantations that would have cleared the higher ridges along the river. By the time of the Civil War, Madison Parish was producing 110,000 bales of cotton per year. After the Civil War, cotton production plummeted and trees began to reclaim the fields.[1]

Early History of Ivory-Bills in the Singer Tract

Sometime before 1916, the Singer Manufacturing Company, makers of Singer sewing machines, bought title to 81,000 acres of land along the Tensas River near Tallulah, in Madison Parish. Some of the land was virgin hardwood forest; much of it was not. Many of the large trees in what came to be known as the Singer Tract in the 1930s could be dated to the immediate post–Civil War era.[2] Singer saw in the forest a continuing supply of quality wood for the cabinets of their popular sewing machines. The area was rich in wildlife, including black bears (*Ursus americana*), red wolves (*Canis rufus*), eastern cougars (*Felis concolor*), and an abundance of white-tailed deer (*Odocoileus virginianus*) and wild turkeys. Local hunters had long favored the bottomland forest of the Singer Tract, and some knew that within this forest was a population of a giant woodpecker, known locally as "kint" because of the sound of its call. Most ornithologists were unaware of the presence of these birds: ivory-billed woodpeckers.

Naturalist E. A. McIlhenny, of Avery Island, Louisiana, knew that the area supported a small population of ivory-bills. In 1925, he and other Louisiana conservationists persuaded the Louisiana Department of Conservation to lease wildlife rights to the area.[3] In 1926, a refuge was established with a ten-year lease on the land, with the Singer Manufacturing Company reserving the right to develop its property or to eventually cut the timber. As a result of the refuge, the birds were somewhat protected; only an occasional poacher shot one for its skin. The habitat, however, was not protected.

When Louisiana legislator Mason Spencer shot a male ivory-bill in April 1932 and presented it to the Louisiana Department of Conservation to prove that they existed, knowledge of the discovery became national news. Audubon president T. Gilbert Pearson and his director of sanctuaries, Ernest G. Holt (who was from Louisiana), visited the Singer Tract in May and argued for giving the birds better protection.[4] Following Pearson and Holt's visit to the Singer Tract, Pearson received assurances from Armand P. Daspit, director of the Fur and Wildlife

Division of the state Department of Conservation, that no additional permits would be issued for collecting ivory-bills. Daspit also indicated that a person had been hired by the Department of Conservation to protect the birds. The state placed two wardens on patrol, and Singer also kept two wardens on duty. But time was running out on the ten-year lease; it would expire in November 1936, and Singer seemed determined to sell the timber.[5]

In June 1933, George Lowery and John S. Campbell spent a week in the Singer Tract searching for ivory-bills.[6] They failed to find them. On Christmas morning 1933, Lowery and Campbell returned with Lowery's father and Jack Kuhn to conduct a Christmas Bird Count in the Singer Tract. In spite of rain and early frustration, they found four ivory-bills, two males and two females, feeding on the same dead snag.[7] The men watched from fifty feet away as the birds fed for fifteen minutes and then flew deeper into the forest.

Study and a Plan for Conservation

At the American Ornithologists' Union meeting in Chicago on October 23, 1934, Lowery presented a paper on the discovery of the ivory-bills in the Singer Tract. He had neither slides nor movies—only the incredible news that ivory-bills still existed. In the audience were the famed ornithologists Arthur Allen, Peter Paul Kellogg, and George Miksch Sutton, and undergraduate student James Tanner. In April 1935, with Audubon's encouragement, Allen, Kellogg, Sutton, and Tanner studied the ivory-bills in the Singer Tract for eight days, as was documented in the previous chapter. At the same time, Allen was laying plans for more work.

Allen sought funds to support basic research on the ivory-bill and looked to wildlife biologist and conservationist Aldo Leopold and the Game Institute for help. Leopold responded in 1936 by sending Allen a draft "Proposal for a Conservation Inventory of Threatened Species." In his proposal, Leopold eloquently defined the problems—problems that persist today—and laid out the rationale for a national park to protect the Singer Tract and the ivory-bill, a proposal that Tanner later championed:

[C]ertain ornithologists have discovered a remnant of the Ivory-billed Woodpecker—a bird inextricably interwoven with our pioneer tradition— the very spirit of that "dark and bloody ground" which has become the locus of the national culture. It is known that the Ivory-bill requires as its habitat large stretches of virgin hardwood. The present remnant lives in such a for- est, owned and held by an industry as reserve stumpage. Cutting may begin, and the Ivory-bill may be done for, at any moment. The Park Service has or can get funds to buy virgin forests, but it does not know of the Ivory-bill or its predicament. It is absorbed in the intricate problem of accommodat- ing the public which is mobbing its parks. When it buys a new park, it is likely to do so in some "scenic" spot, with the general objective of perpetu- ating some definite thing to visit. Its wild life program is befogged with the abstract concept of inviolate sanctuary. Is it not time to establish particular parks (or equivalent) for particular "natural wonders" like the Ivory-bill? You may say, of course, that one rare bird is no park project—that the Biological Survey should buy a refuge, or the Forest Service a National Forest, to take care of the situation. Whereupon the question bounces back: the Survey has only duck money; the Forest Service would have to cut. But is there anything to prevent the three possible agencies concerned from get- ting together and agreeing whose job this is, and while they are at it, a thou- sand other jobs of like character? And how much each would cost? And just what needs to be done in each case? And can anyone doubt that the public, through Congress, would support such a program? Well—this is what I mean by an "Inventory and Plan."[8]

Cutting and the Fight for the Singer Tract Begin

In November 1936, the state's lease on the Singer Tract expired, and in early 1937, just as Tanner was beginning his studies of ivory-bills, the Singer Manufacturing Company entered into a contract with the Tendall Lumber Company, selling them 6,000 acres around Horseshoe Lake and Lake Despair. In 1939, the remaining 74,000 acres were sold to the Chicago Mill and Lumber Company.[9] The lumber companies would cut timber on the tract, and as it was cut, they would take title to the land. Once they had title to the land, they sold it as quickly as possible, often in plots of twenty to sixty acres, to be used as farmland. The going price for the land was about $40 per acre. For legal reasons, until the land was sold—and technically perhaps even after—it remained as a state game refuge. However, enforcement of game laws on the land was next to impossible. Tanner's studies took on great significance as time was running out for these few ivory-bills.

In the spring of 1940, following Tanner's departure from the area, McIlhenny reported that a survey of lands being cut over revealed no ivory-bills.[10] At most, seven ivory-bills were in the remaining forest, and some of those observations might have been of the same birds. Something had to be done. Tanner fought desperately to save the remaining forests. At the annual meeting of the Tennessee Academy of Science, he got a resolution passed urging the establishment of a Tensas National Park in Tensas Parish. He was also successful in getting Senate Bill 329 introduced into Congress by Louisiana senator Allen J. Ellender to permit the purchase of lands for the National Park, although the bill never made it out of committee.[11] The American Ornithologists' Union's Committee on Bird Protection also joined in the plea, urging the state and federal government to purchase the Singer Tract, calling it the ivory-billed woodpecker's "last important habitat."[12] The Bureau of the Budget, however, concerned about needed defense spending, recommended against the measure, and Congress took no action. In August 1941, George H. Bick, a biologist for the state, estimated 25,000 acres had been cut, and the "Chicago Mill and Lumber Co. does not seem inclined to renew the cooperative refuge agreement with the state when it expires in 1946."[13] The cutover Chicago Mill lands were then valued at $10 to $25 per acre, and the cutover Tendall lands were going for $7 per acre, less than half the price of the forested lands just two years earlier.

On December 27, 1941, Tanner was back in the Singer Tract. Early that morning he went to the Chicago Mill and Lumber Company Field Office and found a supervisor, Mr. Alexander, who accompanied him into the field. Tanner, hoping to win a friend for the ivory-bill, showed him the characteristic feeding work of ivory-bills—slabs of bark removed to get at the beetle larvae. Alexander was interested and cooperative, but nothing came of the effort. Tanner noted that Alexander's most pointed comment was, "They ought to learn to feed on something different."[14]

During 1941 and early 1942, the National Audubon Society had been negotiating with the Chicago Mill and Lumber Company, but company president McClelland stood firm in opposition to the society. When McClelland died in February 1942 and a new president, J. F. Griswold, was appointed, negotiations were reopened. In March 1942, Griswold met with John Baker at Audubon headquarters in New York. Griswold showed signs of willingness to cooperate, so Baker gave him copies of Tanner's report from his December visit to the Singer Tract—after tactfully deleting a few passages. Allen believed that the Chicago Mill and Lumber Company felt that the final decision on the property should be made by the Singer Manufacturing Company and was hoping that Singer would release the Chicago Mill and Lumber Company from their contract. The Greenlea Bend area of the Singer Tract, including Section 5, the west half of Section 4, and all of Section 32 to the south of the Tensas River were being considered, but Baker *and* Griswold thought that such a refuge would be too small. So they proposed adding all of the lands east of Sections 34 and 3. (Note that a section is always one square mile.) The proposed refuge would include a little more than 6 1/2 square miles of bottomland forest sandwiched between bends and straddling the Tensas River. It was not the best of ivory-bill habitat, but it was what was left. It probably remained because it was lower and more subject to flooding, thus less accessible to the timber cutters. That would have meant that it was ecologically different as well, much of it not the higher "second bottom" forest preferred by the birds. At best, a refuge of 6 1/2 square miles could support one pair of ivory-billed woodpeckers, not providing much hope for a future for the species. Establishing the refuge was going to be complex, mainly because the various tracts had been valued differently depending on the type, amount, and quality of timber present. At an average of $50 to $60 per acre, Baker told

Allen that they were looking at a purchase price of between $175,000 and $250,000.[15]

With the 1942 publication of Tanner's monograph on the ivory-bill as the first National Audubon Society Research Report, one might have expected a surge of renewed interest in and efforts on behalf of the ivory-bill. Baker wrote an enticing review of Tanner's research report, urging readers to get a copy and emphasizing that neither clear-cutting nor selective cutting was acceptable in providing for the ivory-bill.[16] Baker added, "Whatever action is taken, rest assured that we are doing everything in our power to save the Ivory-bill." There was some response, but not the necessary outpouring of support. Even with Tanner's work, Baker's comments, serious pleas from the National Audubon Society, and support from the National Parks Association, the American Forestry Association, and the American Nature Association, the cutting did not slow.[17] As World War II progressed, there was increased demand for hardwood timber for planking for the decks of PT-boats, trucks, and especially for pallets to hold the ammunition and other equipment being shipped overseas. Wood was described by the government as our "most critical raw material production problem" and a major focus was on southeastern forests.

Conservation-minded citizens recognized the urgency of the situation. Henry B. Chase Jr. of New Orleans, in a letter to the editor of *Nature Magazine,* pleaded, "It is most imperative that all scientific organizations and individuals use every means at their disposal in trying to save at least part" of the remaining Singer Tract acreage.[18] As a result of this type of grassroots support and continued lobbying by Baker, Tanner, and others, Senator Ellender again introduced a bill to establish the Tensas Swamp National Park.[19] The bill called for private donations of funds—less than $2 million was needed for the then estimated 60,000 acres. To give impetus to the bill, Rosalie Edge, conservation activist and critic of the National Audubon Society, through her Emergency Conservation Committee, urged the National Audubon Society to do its part by donating or guaranteeing "a generous sum."[20] Again, the bill failed to make it out of committee. Late in 1942, Baker appealed directly to President Franklin D. Roosevelt, and Roosevelt responded with interest, sending his reply to Baker to Secretary of the Interior Harold L. Ickes. Baker contacted Ickes; Ira Gabrielson, chief of the US Fish and Wildlife Service; Newton B. Drury, director of the National Park Service; and Earle H. Clapp, acting chief of the US Forest Service. Clapp told Baker he did not think it would be necessary to cut all the old-growth timber in the Singer Tract. Baker pressed on and the head of the Lumber Division of the War Production Board told him that, pending verification of data, some of the Singer Tract might be excluded from the war production effort.[21]

The swamp forest of the Singer Tract, however, did not have the park status appeal of other national parks, and support for the bill was weak. Arthur Newton Pack, editor of *Nature Magazine,* wrote an editorial summarizing the Park Service's and Audubon's efforts and supporting preservation of the Singer Tract as a national monument "preserving a typical area of outstanding biological importance."[22] He also noted Senator Ellender's bill to create the Tensas Swamp National Park that would preserve "an area not exceeding 60,000 acres," but fell short in his endorsement by commenting, "While the region hardly fulfills requirements for National Park designation, it certainly should be saved."

Not everyone was convinced that strong enough efforts were underway. In 1943, Edge added her voice to the discussion of preserving the Singer Tract as a National Park, although noting that it is "now almost completely destroyed." She lamented, "We cannot with honesty urge that the purchase of the Singer Tract will save the Ivory-billed Woodpecker, whose last stand is now made there. Had the forest been protected sooner, the Ivory-bill might have been saved. But continuing cutting has narrowed its habitat, and probably only six Ivory-bills now exist on the Singer Tract."[23]

Edge placed considerable blame for the failure to protect the Singer Tract on the National Audubon Society, calling the Audubon Society's efforts to save the Singer Tract "another example of action which comes too late." She recounted Baker's meetings and correspondence with

Singer and Washington officials and mocked his efforts: "'Rest assured,' he says, 'that we are doing everything in our power to save the Ivory-bill.'" To the latter she responded,

> Save the Ivory-bill? It is too late, too late. The National Audubon Society was warned again and again of the danger to the Ivory-bill, particularly in our pamphlet, *Crisis in Conservation*, signed by two curators of the American Museum of Natural History, and distributed to all Audubon members in 1931. To-day, the death knell of the Ivory-bill is tolling. And instead of the living bird, we have offered to us a handsome book on the Ivory-bill (price $2.50), published by the National Audubon Society—a sad commentary on "research" in lieu of protection.[24]

Additional attacks on the National Audubon Society followed, suggesting that had the money given to Tanner been used as seed money to raise funds to buy the Singer Tract, the ivory-bill might have survived. Recognizing that "saving" the Singer Tract in 1943 was not likely to save the ivory-bill but might save a fragment of a unique ecosystem, Edge challenged Audubon to guarantee the private funds required in Senator Ellender's bill for purchase of the area. She said that there was nothing that her Emergency Conservation Committee would like more than to be able to contribute to such a fund. Her challenge was unanswered.

The US Forest Service was asked to oversee and coordinate the timber industry in the war effort, and in August 1943, the War Production Board established the Timber Production War Project—abbreviated TPWP and commonly known as "TeePee WeePee."[25] The result was that the rate of cutting in the Singer Tract increased dramatically, aided by the use of German prisoners of war. All suggestions by government officials that some of the old growth of the Singer Tract might be spared seem to have been forgotten. By November 1943, only about 12,000 acres of primeval forest suitable for the ivory-bills remained.

In November 1943, Baker made another pilgrimage to the Singer Tract, and he found that only fifteen square miles of virgin hardwood forest remained uncut, including an area that ivory-bills had been using. He only saw a lone ivory-bill. Reporting back to the Audubon membership, he stressed that even without the ivory-bill, this forest was worth saving. While in Louisiana, Baker had an interview with Gov. Sam H. Jones and conservation commissioner Joseph L. McHugh, and they pledged $200,000 to buy as much of the remaining old-growth forest as they could so that it could be set aside as a wildlife refuge.[26]

The governors of Louisiana, Tennessee, Arkansas, and Mississippi added their support in a joint letter, dated November 22, 1943, to Chicago Mill and Lumber Company and to Singer Manufacturing Company:

> We the Governors of Louisiana, Tennessee, Arkansas and Mississippi request that the Chicago Mill & Lumber Co., agree to now waive its contract rights to cut the remaining virgin timber in the so-called Singer Tract near Tallulah, Louisiana, and agree to sell at a reasonable price for inviolate wildlife refuge purposes sections of land in a cutover buffer area.
>
> We also request that the Singer Mfg. Co. agree, in the event of waiver of rights by the Chicago Mill & Lumber Co., to sell for inviolate wildlife refuge purposes the sections containing virgin timber at prices no higher than those contained in its current contract with Chicago Mill & Lumber Co. We feel that the saving for all time of the rarest North American bird, the Ivory-billed Woodpecker, and of the finest remaining stand of virgin hardwood timber in the South are matters of great national importance, and of concern to the citizens of our states.
>
> Signed: Sam Jones, Louisiana; Prentice Cooper, Tennessee; Homer Adkins, Arkansas; Paul B. Johnson, Mississippi.[27]

At the instigation of Governor Jones, a meeting was held in early December 1943 at the Chicago Mill and Lumber Company offices in Chicago to discuss the possible purchase.[28] Those present included the chairman of the board and president of the lumber company; Commissioner McHugh from Louisiana; John Clark Salyer II, chief of the Division of Wildlife Refuges of the US Fish and Wildlife Service; a federal attorney; and Baker, representing the National Audubon Society. Louisiana wanted to buy forest and the federal government was prepared to lease, with options to buy, a buffer area of cutover land around the preserved forest. Governor Cooper of Tennessee expressed strong interest in saving the remaining forest and offered to see that his state appropriated funds if a collaborative effort could result in saving the Singer Tract.[29]

The answer was not just "no"; it was an emphatic "no!" The lumber company was unwilling to discuss any possibility of cooperation in setting up a wildlife refuge for the ivory-bill if it meant any limitation whatsoever on its contract rights to cut timber.

World War II was foremost on everyone's mind. And the National Audubon Society also had other priorities, but Baker tried again—this time by using the press. On February 22, 1944, he made an impassioned plea to establish a wildlife refuge on the remaining Singer Tract. His message was delivered before the convention of the Outdoor Writers Association of America in Columbus, Ohio.[30] Baker related the history of the Singer Tract, the rarity of the ivory-bill, and the efforts made at the meeting in Chicago. He also pointed out that conservationists should not single out Chicago Mill and Lumber Company for chastisement, because they were no different from many other companies. None of them, he said, made any reforestation efforts or did any selective cutting. Essentially, he said, the forest industry "is bent upon 'getting it while the going is good,' and seemingly cares not for the future."

Fading Hopes

In anticipation of possibly saving what was left of the Singer Tract, Baker sent Richard H. Pough, a conservation biologist working with threatened species for the National Audubon Society, to the area before the Chicago meeting.[31] Pough stayed at a hunting camp on the privately owned Sharkey Plantation on the Tensas River just south of the Singer Tract. He spent long days in the Singer Tract assessing the nature and availability of habitat for the ivory-bill, the rate at which habitat was being lost, and searching for the birds. He found a lone female in the John's Bayou area in the northeast quarter of Section 61 south and east of the Bayou. He followed her for two full days and noted that she foraged on relatively small Nuttall oaks that appeared to be dying as a result of being shaded out by taller trees. During that time the ivory-bill never once entered the adjacent cutover area.

Otherwise, most of Pough's time in the Singer Tract was spent in the Mack's Bayou area. He noted that it was the only extensive area left in which sweetgums were fairly common. Although there were many sweetgums up to five feet in diameter, they were scattered across the forest rather than located in old stands.

When Pough left the Singer Tract on January 19, 1944, his recommendation to the National Audubon Society was to acquire about ten square miles of intact and cutover forest centered on the area of previous proposals. He saw in this area a mosaic of habitats and the potential for good sweetgum habitat. Pough's recommendation to the National Audubon Society was to acquire forest in Sections 23 to 26 and 34 to 36, Township 15 N, Range 11 E. Pough also gave recommendations for purchase of additional habitat to include part of the Sharkey Plantation and lands along the Tensas River. For buffer areas, he suggested purchasing cutover lands in the Titepaper Ridge, John's Bayou, and Greenlea Bend areas.

Pough ended his report with a pessimistic prediction for the ivory-bill's survival but recognized that in losing the Singer Tract we were losing much more than just the ivory-bill:

> While I doubt whether the preservation of this area would save the Ivory-billed Woodpecker, it would preserve an area of quite typical Mississippi bottomland hardwood forest, with all its beautiful and luxuriant vegetation, and all its original wildlife except the Ivory-billed Woodpecker and the Louisiana Paraquet [*Conuropsis carolinensis*]. If any Ivory-bills still survive in the neighborhood, such a refuge unit would at least give them some place to go and at least some chance to survive.[32]

Pough also commented about his observations of the lone ivory-bill, suggesting the bird was specifically avoiding the cutover area.

Tanner corresponded with Pough, stating that when he had studied two females in the same area during 1941, "one of them roosted in a dead oak snag containing three large holes, the snag standing within 100 feet of the section line. . . . during that time . . . the two birds did quite a bit of feeding in the cut-over."[33]

Tanner also noted that the disparity in their observations fit a pattern he had witnessed in 1939 when the Hunter's Bend area of the Singer Tract had just been cut over and also in other areas that had been cut over less than a year. Tanner observed that most species of woodpeckers were "quite common," whereas in areas cut over about eighteen months earlier, there were few woodpeckers. I have seen the same pattern in South Carolina, Mississippi, and elsewhere in Louisiana. The debris from cut trees attracts a lot of bark beetles, whose populations grow rapidly in response to the freshly cut slash. The stumps and cut limbs of the harvested trees give off chemical signals that attract the insects and provide food and conditions perfect for their reproduction. Beetle numbers in turn attract the woodpeckers. As the slash from the cutting deteriorates, the quality of habitats for the insects declines quickly and beetle populations suffer both from predation by woodpeckers and a lack of a food supply. Then, lacking an abundant food supply, the opportunistic woodpeckers move to more favorable areas.

Following the failed meeting in Chicago and Pough's report, Baker sent yet another prominent ornithologist to the Singer Tract in the spring of 1944. This time it was Ludlow Griscom who was to evaluate prospects for the birds. Griscom was then both the chair of Audubon's board of directors and a distinguished researcher at Harvard's Museum of Comparative Zoology. His charge was to see what chance there was of saving the last great stand of virgin baldcypress and any ivory-bills that might still live there. Griscom reported back to Baker at Audubon and Thomas Barbour at Harvard. Barbour's response reveals the state of things at the Singer Tract: "I awaited his report fearfully, and was not surprised when he said: 'The whole area is full of portable sawmills. You can't get away from the sound of tractors hauling out logs. It is too late. It can't be done.'"[34]

In April 1944, shortly after Griscom's visit, wildlife artist Donald Eckelberry went to the Singer Tract under contract to the National Audubon Society with the hope of providing one last artist's perspective of the bird and its habitat. He spent nearly two weeks and observed a lone female in the Singer Tract.[35] He sketched the bird, creating perhaps the last paintings of an ivory-bill sketched from life (Figure 8.1).

A Refuge, But Too Late for the Ivory-Bill

In December 1948, one of Tanner's former students, Arthur Mac-Murray, visited the Singer Tract in hopes of finding ivory-bills. Tanner had provided MacMurray with contact information and eagerly anticipated his report. On January 8, 1949, MacMurray wrote to Tanner, and his letter provides scant hints that the ivory-bill might have survived at least until late 1948—but the news was not good:

> Singer Tract has been cleaned of all its commercial lumber as far as I could gather. No ivory-bills have been seen at John's Bayou for 3 years, according to a resident who has lived adjacent to it for the past 22 years. . . . John's Bayou has a lumber railway passing through it and passing all the way north to some point due west of Tallullah. The ivory-bills left John's Bayou soon after the large gum tree which had been their nest tree for several years was lumbered.[36]

McMurray noted that only one pair of ivory-bills was believed to be in the region, having been seen in November 1948 near "North Lake #1" and that they were "apparently wandering over much larger areas than formerly." The last stands of old sweetgum trees in the Singer Tract were then being cut. MacMurray reported what may have been the last sighting of ivory-bills in the region, although the report was secondhand: "A friend of the gentleman who resides adjacent to John's Bayou reported that he saw what he thought was an ivory-bill on E. C. McCallip's property on the Little Fork Road 6 miles south of Waverly on December 17th of 1948."[37] With high hopes, McMurray visited

Figure 8.1. Donald Eckelberry's painting of perhaps the last ivory-bill in North America, a female, flying across the cutover forest of the Singer Tract in April 1944. *Courtesy Jerome A. Jackson*

McCallip's property on December 24, 1948. It was cut over. He saw a lot of woodpeckers, but no ivory-bills.

John Dennis, a prominent woodpecker biologist who found ivory-bills in Cuba, also suggested that the last ivory-bill seen in the Singer Tract was in 1948, although he provides no details as to specifically when, the circumstances, or the observer.[38] We are left to wonder if those last birds died in the Singer Tract, or if in their wanderings in an ever widening search for suitable habitat they might have taken refuge in remaining forest along the Mississippi, to the south along the Tensas River, or in the Atchafalaya Basin.

Helen Ossa in her book *They Saved Our Birds* wrote regarding the Singer Tract,

> while it was being heavily logged, conservationists succeeded in persuading President Franklin D. Roosevelt to have the area in which the ivory-billed was last seen set aside as a sanctuary while an investigation of the possibility that the bird still lived there, was made. As is the way with this elusive bird not a one could be found, so the logging was resumed and the area is now completely logged over.[39]

I have found no other reference to Roosevelt having stopped logging in the Singer Tract or set aside the Singer Tract as a sanctuary, although pleas from the National Audubon Society had made it to the president's desk.[40] Certainly the greater share of the logging had been completed by 1944 when Eckelberry and others had visited.

Long after the ancient forest of the Singer Tract was gone and the haunting notes of ivory-bills were only a memory, the Singer Tract finally became a national wildlife refuge. Public Law Number 96-285 was approved on June 28, 1980, directing the secretary of the Interior and the secretary of the Army to acquire a refuge including about 50,000 acres in Tensas, Madison, and Franklin Parishes, Louisiana. The law provided for $40 million in purchase authorizations for the Army and $10 million for the Department of the Interior. Lands acquired by the Army were to be turned over to the Fish and Wildlife Service for management. Cutting was still going on in the area, however, and Chicago Mill and Lumber Company was still much involved. Controversy continued to reign. This time, however late, the lands were secured and the cutting stopped.

In 1985, Public Law Number 99-191 provided another $10 million to the Department of the Interior for additional Tensas Refuge lands.

Tanner visited the Singer Tract in 1986 as plans were being made for a new visitor center at the refuge. He walked through the woods with refuge manager H. T. "Tuck" Stone, reminiscing about the forest that had been and the birds that Tanner followed day after day. On his return to Tennessee, Tanner sent sixty-nine negatives he had made of ivory-bills and their habitat between 1935 and 1941 to the refuge for use in displays documenting what once existed.[41]

Finally, on June 25, 1988, the Tensas River National Wildlife Refuge Office and Visitor Center was dedicated. Sen. John Breaux presided over the festivities. Gov. Charles E. "Buddy" Roemer III, Rep. Jerry Huckaby, and regional director of the US Fish and Wildlife Service James Pulliam spoke.[42] So did several state and local dignitaries. Tanner was invited as a special guest but sent his regrets. There were two other special guests, however—two ivory-billed woodpeckers, specimens lacking data but possibly collected from the Singer Tract. They had been donated for an exhibit at the refuge by Cornell University.

CHAPTER 9

In the Footsteps of Others

A light car can be driven to the edge of swamp in dry weather, but inside the swamp legwork is needed and a pair of webbed feet would be useful.

> —James T. Tanner, from the text of a report presented to the National Association of Audubon Societies regarding his search for the ivory-billed woodpecker, October 1939

For more than a century, people from all walks of life have searched for the ivory-billed woodpecker. Among the early searchers were ornithologists, hobbyists, and mercenaries. Before 1900, all known ivory-bill seekers were hoping to return with the skin or eggs of their quarry. Today seekers hope only to bring back photos, videos, or sound recordings—some tangible proof that the ivory-bill still flies in some remote forest.

Some searchers have been successful. Still others, believing they were successful, have been viewed skeptically by the scientific community. Others have admitted defeat but reveled in the experience—most eager to follow the next lead. Ornithologists Arthur Allen and James

Tanner have been the most successful ivory-bill searchers. But others have included vice president of General Mills Whitney Eastman; physician George Reynard; and lawyer Hal Owens. Julie Zickefoose, Christopher Cokinos, and I have previously reviewed some of these searches, chronicling the difficulties and lure of the hunt for this bird that is sometimes called "the holy grail of birding."[1]

These searchers were my immediate predecessors and contemporaries. Evaluating their efforts and learning the history of ivory-bills in each area was essential for developing an understanding of the problems in searching for this bird and in planning my own searches. The nature and results of my searches in North America are interwoven with these other efforts, putting them into the context of our hopes for the ivory-bill during the last half of the twentieth and early twenty-first century.

More recent searches have benefited from technology: GPS (global positioning system) units can keep one from getting lost and cell telephones can keep one in touch in case of emergency or to allow coordination of efforts. But I had neither of these in the 1980s. Legwork, dealing with a mosaic of water, soggy land, and dense thickets of briars—in addition to mosquitoes, ticks, poisonous snakes, and hunters—was, and is, the order of the day if one is serious about looking for ivory-bills in the southeastern United States.

In 1985 and 1986 I identified some specific areas to be searched for ivory-billed woodpeckers through discussions at Ivory-Billed Woodpecker Advisory Committee meetings. Others were selected because of apparently suitable habitat in an area relatively close to sites of confirmed records during this century; others because of topographical maps or satellite photos showing or suggesting an extensive bottomland forest. I did not visit some sites that we discussed or visited them only briefly because I learned that they had been the subject of searches by other biologists in recent years and because reports and results suggested that efforts elsewhere were more important.

Publicity concerning the search for ivory-billed woodpeckers generated hundreds of responses from individuals who believed they had seen the bird. Of these, approximately twenty reports provided details that could not easily be discounted, and these provided additional targets for habitat evaluation and field searches. What started out as a two-year effort was extended into a four-year effort—with no additional funds. Part of the reason for the extension was time taken out to go to Cuba as a member of the Piciform Specialist Group of the International Council for Bird Preservation. Part of the reason was simply the morass of logistics.

My searches for potential ivory-billed woodpecker habitat in the United States have taken me to Texas, Louisiana, Arkansas, Missouri, Mississippi, Alabama, Georgia, South Carolina, and Florida. Preparation for each site visit included review of the ivory-billed woodpecker literature for the area, review of specimen data, correspondence with individuals in the area or familiar with the area, maps or aerial photographs, and selection of a base of operations.

Ground visits began by contacting land managers and driving back roads into the area. Where appropriate and possible, I made an aerial reconnaissance using a small plane to further evaluate the extent and nature of the habitat and to assist with selection of areas to visit on foot or by boat. I was fortunate in that regard because I am a pilot and was familiar with low-altitude flying over forested areas. I had previously developed aerial survey techniques for locating red-cockaded woodpecker cavity trees.[2] I knew I could do a fair evaluation of habitat from the air and might even be able to spot flying ivory-bills because of their size, distinctive black-and-white pattern, and reported habit of flying above the tree tops when going long distances.

In general, for ground efforts I followed the approach used by Tanner, trying to cover an area widely first rather than thoroughly, to be able to judge the nature of the habitat available.[3] In some cases such a superficial evaluation was adequate for me to judge the site as not having a reasonable probability of being able to support ivory-billed woodpeckers, and I made no more efforts. I considered the best areas to be those that

1. included a large expanse (3,700+ acres) of relatively mature, inaccessible bottomland forest in which large (twenty-four- to thirty-six-inch diameter at four feet above ground) sweetgums and oaks could be found readily;
2. were within about forty miles of a site at which Ivory-bills were known during the twentieth century; and
3. were contiguous with or within a few miles of additional similar habitat.

When I reached potentially suitable old-growth habitat, my normal procedure was to hike/wade/canoe for approximately fifteen minutes, then to play forty-five seconds of the Cornell recording of a single ivory-bill giving single and double *kent* notes. I then listened for three minutes for any response, noted habitat characteristics (tree species, nature of the understory, etc.), took several basic forest measurements, and recorded the presence of other woodpecker species heard or seen. I took habitat measurements only in what appeared to be the best available habitat. My original intent was to take measurements in all forest habitats in a given area, but this was not feasible given the time and resources available. At all times I was listening for the characteristic notes of the birds and looking for feeding signs (bark scaled from recently dead trees and limbs for a considerable extent) and for potential ivory-bill cavities. I also carried with me unmarked illustrations of ivory-billed, pileated, and red-headed woodpeckers that I showed during discussions with interested individuals.

Such a widespread search as I was making resulted in a number of logistical problems that hampered my efforts. Among them were gaining access to private land, activities of hunters, flooding, and bad weather.

Although there are large areas of public land that include potential ivory-billed woodpecker habitat, many of the areas I visited were private lands, and finding ownership and gaining permission to enter the lands was time-consuming—and sometimes impossible. In the latter cases, I limited efforts to public lands and roads, streams traversing private lands, or to aerial reconnaissance.

Hunting seasons in the Southeast are long—and dangerous for a person trying to move inconspicuously through a large forest tract. Some deer seasons were open as early as August (South Carolina) and others remained open through January (Mississippi)—only to be followed by the opening of turkey season in early spring. Although I made several fall trips into potential ivory-bill areas, these were generally limited because of hunters. Some lands that I searched were leased by hunt clubs (e.g., along the Yazoo, Pearl, Tombigbee, Noxubee, and Pascagoula Rivers in Mississippi), and permission to enter these lands usually could only be obtained outside of hunting seasons.

It often seemed as if the hunting seasons were timed to coincide with the dry season and ended when the rains began. Spring—when I made most of my trips—often meant high water and difficult access.

Because of teaching duties during fall semesters and during the spring of 1989, my searches were then limited to four-day weekends. Rainy weather often impeded otherwise well-planned searches, and I got used to being absolutely soaked.

I have organized this narrative of my efforts and discussion of other efforts alphabetically by state to provide inclusive review of areas of potential habitat. Some localities that I visited no longer seem to provide suitable ivory-billed woodpecker habitat and are discussed briefly in Appendix 1. Rivers often form state boundaries, thus some prime riverine swamp habitats extend across state boundaries (e.g., Pearl River Swamp in Mississippi and Louisiana), in which case, my discussion is focused in one state account and cross-referenced in the other.

Arkansas

The last Arkansas population of ivory-billed woodpeckers is believed to have existed near the mouth of the Arkansas River and to have disappeared between 1900 and 1915.[4] However, ornithologist Douglas James received a report of ivory-bills in Jackson County in 1986.[5] Two groups of Arkansas birders visited the area in 1986 and 1987, with negative results.[6] I visited southeastern Arkansas in 1986 and 1988 and focused on habitats along the Mississippi and lower Arkansas and White Rivers. I drove available roads in suitable habitat between Jonesville-Blytheville and the Louisiana line, staying within the alluvial plain of the Mississippi. Superficial examination of habitats revealed few areas of extensive, mature bottomland hardwoods, and I discontinued my efforts in the state.

Florida

Ivory-bills were reported in the Chipola River Swamp in 1950 by Eastman and Muriel Kelso, and also apparently in the same area by Davis Crompton.[7] Through the combined efforts of the National Audubon Society, the Florida Game and Fresh Water Fish Commission, local landowners, the St. Joe Paper Company, the Neal Lumber and Manufacturing Company, and Kelso, the 1,300-acre Chipola River Wildlife Sanctuary was established to protect any ivory-bills that might be there.[8] In the absence of additional sightings, the sanctuary status was discontinued in 1952.[9]

Tanner told me that he spent several days in 1950 around the Chipola River and came to the conclusion, reached also by wildlife biologist Herbert Stoddard, who had visited the area, that there was no evidence of ivory-bills. John Dennis reported hearing an ivory-bill call five times in the Chipola Swamp on April 5, 1951, apparently the last report of the species from the area.[10]

In 1952, Richard H. Pough, then secretary of the US Section of the International Committee for Bird Preservation, wrote,

> The cutting of the last remaining virgin swamp forests of the southern states which is now in progress seems likely to result in the ultimate extinction of the magnificent Ivory-billed Woodpecker. . . . A few still survive in the Apalachicola River delta of western Florida, but no plans have yet been made to preserve permanently this or any other sizable block of southern swamp forest, which might give the Ivory-bill and the other unique wildlife of these swamps a chance for survival.[11]

In the summer of 1967, ornithologist David Lee, then a university student, saw what he identified as a female ivory-bill fly across the road about twenty-five yards in front of him as he was driving north of Haines City where the road cuts through the Green Swamp in central Florida.[12] The white at the back of the wings and absence of red on the head clearly identify a female ivory-bill. Lee and his professor, John Funderburg, returned to the area, found a grove of pines that had been burned by a lightning-started fire, and discovered that large slabs of bark had been removed from several of the trees—a characteristic of ivory-bill feeding activity.

In the late 1960s, another pair of ivory-bill hunters, H. Norton Agey and George M. Heinzmann, were the subject of news stories and birding hopes. The central Florida searches of Agey and Heinzmann included forty-one days spent in an area where they reported having seen or heard an ivory-bill on eleven occasions between 1967 and 1969.[13] Dennis identified the area as being in Polk County, northwest of Lake Okeechobee, but he later said that the area was in Highlands County (also northwest of Lake Okeechobee).[14] The only tangible proof of their observation was the innermost secondary feather of an ivory-billed woodpecker (plus some body feathers that might have been from an ivory-bill), which they reported, found near a cavity that had broken open when the tree fell. Alexander Wetmore of the Smithsonian Institution confirmed the identity of the feather. I examined copies of the correspondence between Wetmore and these authors and the feathers found; they are with the bird collection at the Florida Museum of Natural History in Gainesville.[15] The feather is indeed one of an ivory-bill. Although the length of time it had been in the cavity could not be determined, the secondary feather is in good shape. I accept these two statements as valid on the authority of Dr. Wetmore. However, some shadow of doubt is cast over these records because Agey and Heinzmann also tape-recorded what they said were ivory-billed woodpeckers, and personnel at the Cornell Laboratory of Ornithology identified the birds on those tapes as pileated woodpeckers.[16]

In the Big Cypress area, William Robertson reported that Allan Cruickshank, a prominent ornithologist, saw an ivory-bill fly across Florida Highway 29 in about 1950.[17] I have not been able to obtain additional details of this report, but Cruickshank was a credible observer. Samuel Grimes, another competent Florida birder, reported seeing an ivory-bill in July 1952 about twenty miles south of Tallahassee.[18] John K. Terres, former editor of *Audubon* magazine, reported seeing two ivory-bills on April 9, 1955, south of Homosassa Springs, Florida, but provided no details.[19]

In August 1966, Bedford P. Brown Jr. and Jeffrey R. Sanders, Chicago birders, reported watching two ivory-billed woodpeckers as they fed on beetle-infested pines near Eglin Air Force Base. Dennis searched the area and did not find the birds but considered the sighting valid.[20]

Dennis G. Garratt described seeing a male ivory-billed woodpecker along the Loxahatchee River within Jonathan Dickinson State Park on April 30, 1985. His descriptions of plumage and call note clearly suggest an ivory-billed woodpecker.[21] The bird was wary but was seen at twenty-five to forty feet over a period of about fifteen minutes.

On the basis of these reports and discussions with ornithologists and birders familiar with Florida habitats, I spent time searching in several areas, literally spanning the state. In spite of Florida's rapid development, even metropolitan areas are within easy reach of relatively wild areas. But those areas are rapidly shrinking and being altered by diversion of water for human uses and commercial uses of forest products.

About a third of the Big Cypress Swamp—more than 1.2 million acres—is dominated by baldcypress. Even here, however, the loggers won out. Indeed, this vast swamp was one of the last to be logged. Although cutover, in 1974, Congress set aside 568,000+ acres as the Big Cypress National Preserve. Another 98,000+ acres of the swamp are protected in the Fakahatchee Strand State Preserve, 26,400 acres in the Florida Panther National Wildlife Refuge, and 11,000 acres in National Audubon's Corkscrew Swamp Sanctuary. All of this is adjacent to Everglades National Park.

In March 1983, my wife Bette and I conducted helicopter surveys of portions of the Big Cypress Preserve for red-cockaded woodpeckers. At the time we were also specifically aware of the potential for ivory-billed woodpeckers in the area. We saw no habitat that looked exceptional for ivory-bills and, in addition, noted an abundance of trails left by off-road vehicles. From the air, the off-road vehicle trails left the Big Cypress National Preserve looking like a brown and green mosaic that some naughty child had scribbled all over. National Park Service biologists Gary Patterson and Robertson did extensive searches of the Big Cypress area for red-cockaded woodpeckers and would also have immediately recognized an ivory-bill had they seen one.[22] There has also been considerable ecological work associated with the Florida panther (*Felis concolor coryi*) in remote areas of the Big Cypress Preserve and the adjacent Fakahatchee Strand State Preserve, and more recently in the Florida Panther National Wildlife Refuge.[23] Since the 1980s, there have been extensive ecological surveys in the area in conjunction with oil exploration and Everglades restoration efforts. I note, however, that environmental impact assessments I have examined make no mention of even the possibility of ivory-billed woodpeckers occurring in the area.

In spite of my previous experience in the area and current environmental assessment activities, I considered searching for ivory-billed woodpeckers in the Big Cypress Preserve, but I did not as a result of a letter from Robertson in response to my inquiry concerning possible search localities. He stated the following: "I don't think that the Big Cypress region should be very high on your list of priorities. The last report that I know of was around 1950—Allan Cruickshank's sighting of one flying across S[tate] R[oute] 29—and with all the recent activity in Big Cypress and Fakahatchee one would have expected at least a whisper, if there's still anything there."[24]

Although Robertson's thoughts on the Big Cypress were negative, I kept hearing suggestions that maybe ivory-bills could have survived in the Fakahatchee Strand to the west of the Big Cypress National Preserve. I had received two reports of ivory-bills in the Strand in response to news stories about my search, and had briefly visited the area in 1988. When I moved to southwest Florida in 1999, I was determined to spend more time there. Since 1999, I have had three more reports of ivory-bills from the area—all independent of one another—and have made several forays into the area, primarily walking along old railroad beds that were put in to take out the virgin baldcypress. Thus far results of my searchers there have been negative.

The Fakahatchee Strand is a state preserve dominated by sawgrass (*Cladium jamaicense*) and strands of baldcypress and other swamp forest along northeast–southwest oriented low areas. Preserve biologist Mike Owen knows the area about as well as anyone and he holds out some measure of hope; he believes he has had a glimpse of this ghost bird.

My searches in north Florida included the lower Suwannee River where William Brewster and Frank M. Chapman collected a single ivory-bill and saw one other in March of 1890.[25] Their discussions with local residents led them to believe that the bird was then rare in the area. Specimens I have examined and other literature records suggest that as many as 200 ivory-bills had been killed in the general area in the previous twenty years.

My visits to the Suwannee included December 1988, during which time I drove back roads, hiked cross-country through the Jena Wildlife Management Area and into a portion of the California Swamp, and canoed on portions of the Suwannee between Fannin Springs and Fowler Bluff. I often found it necessary (and convenient) to hike along long-abandoned railroad beds similar to those I found in the Fakahatchee Strand—all of which had been constructed to remove the timber early in the twentieth century. In both areas these have modified drainage and vegetative composition. We found a few live oaks (*Quercus virginiana*) and an abundance of cabbage palms large enough to contain an ivory-bill nest cavity, and even found an ivory-bill or pileated-sized cavity that had considerable scaling of bark below it. But tantalizing though it was, the only large woodpeckers I saw or heard were pileated. My efforts in the lower Suwannee River area were concentrated in Dixie and Levy Counties.

Another area that, on the basis of historic records and occasional recent reports, held some promise was the Wekiva River area of central Florida. In researching the area, I learned that Wekiva Springs had originally been called Clay Springs and that in the 1880s railroads had turned it into a tourist destination. The spring waters were thought to cure rheumatism and kidney and bladder ailments. In 1881, Brewster described a boat trip up the Wekiva from the St. John's River and mentions having seen a pair of ivory-billed woodpeckers where "willows and sweetgums took the places of the cypresses next the stream, while in the background palmettos [*Sabal palmetto*] reared their grotesque heads and hummocks of swamp oaks shut out the sky."[26] By the 1920s, however, tourists had found other attractions, and lumbering took over. Through the 1920s and 1930s, basswood (*Tilia americana*) from the region provided wood for cigar boxes. Then in 1935, the Wilson Cypress Company built a narrow-gauge railroad into the swamp to remove the virgin baldcypress.[27]

Today the Wekiva River and its three main tributaries flow through approximately 20,000 acres of state lands that are more or less preserving the natural habitat: Wekiwa Springs State Park, Rock Springs Run State Reserve, and the lower Wekiva River State Reserve. Florida has also protected the river itself by designating it as the Wekiva River Aquatic Preserve and an Outstanding Florida Water, and by designating a portion of the river as scenic and wild. To the north and west, one of the tributaries of the Wekiva, Blackwater Creek, has its headwaters in Ocala National Forest. The lower Wekiva River State Reserve includes approximately 4,640 acres of wilderness containing narrow strands of swamp forest along the lower reaches of both the Wekiva River and Blackwater Creek. Rock Springs Run State Reserve includes about 4,450 acres of swamp forest, plus a larger area of upland forest. The wild character of these reserves is suggested by the population of black bears that they support.

I received two independent reports of possible ivory-bills from along the Wekiva River in 1987 to 1988, neither of which could easily be dismissed as a pileated report. On visiting the area I found Brewster's description equally applicable today—although he did not mention the sizes of trees, and most I found were clearly of the twentieth century. Brewster did not mention the extent of forest beyond the river's edge, but doubtless it was extensive then, and now it is clearly fragmented and limited.

My visits to these areas included December 1987, January and May 1988, and April 1989. I found the upper Wekiva to be a high-use area for canoeists, although their activities are limited to the river itself. For an area that is close to being in metropolitan Orlando, the adjacent swamp forests seemed quite undisturbed. One of the recent ivory-bill reports came from Wekiva Springs Run, a short stretch between the Wekiva Springs in the state park and the confluence with Rock Springs Run where the Wekiva River proper begins.

That report, by Dale Braiman, specifically mentioned the black crest of the woodpecker seen.[28] This is one of the most heavily used portions of the river, yet it is bordered by mature forest that is rarely penetrated by humans except for a short boardwalk and a trail immediately adjacent to the run. Swamp forests in the park and adjacent Rock Springs Run State Reserve include baldcypress, sweetgum, red maple, southern magnolia, live oaks, and cabbage palms, with many trees exceeding thirty-five inches in diameter in relatively undisturbed stands. In the area of the upper Wekiva I was impressed by the abundance of dead trees scattered through the forest, by the tangle of rotting logs on the forest floor, and by the abundance and diversity of woodpeckers.

The area along the lower Wekiva River and Black Creek includes only narrow strands of swamp forest adjacent to the streams. About seven miles upstream there is extensive swamp associated with Blackwater Creek. I got no responses to playback of ivory-billed woodpecker calls, and saw no foraging sign that I could attribute to ivory-bills, but the network of interconnecting waterways, state preserves, and National Forest has some potential for supporting ivory-bills. The biggest arguments against their presence in the area is the presence of professional biologists working in each of the areas and the certain frequent visitation by bird-watchers—if ivory-bills are present, they should have been seen.

It seems unlikely that there is a resident population of ivory-bills in the area, but quite possible that the quality of the habitat would be attractive to an individual dispersing from some remote area less frequented by humans. Unfortunately, the future of this area is clouded by subdivisions that extend to the very edges of the state lands on the south side and by pressures to allow a beltway to cross the Rock Springs Run State Reserve.

The final area in Florida that I searched included bottomland hardwood forest and baldcypress swamp areas along the Chipola and Apalachicola Rivers between Marianna/Chattahoochee and approximately Dalkeith. This area was selected as a result of the work of Eastman and others in the 1950s.[29] Paul Sykes searched areas along the Apalachicola and Chipola in 1967 and found "habitat excellent for ivory-bills" in some areas.[30]

My visits in June and December 1987 and January and October 1988 took me into extensive tracts of bottomland hardwood and mixed swamp/upland forest that are protected by state and federal agencies, although before protection, the area had been extensively cutover. Some forest areas that are maturing perhaps now are suitable habitat for ivory-bills, but they might not have been good habitat in the 1940s and 1950s. Some expanses of forest and bottomland areas provide reason for hope. One such area is west of Abe Springs and south to Coconut Bluff along the lower Chipola in Calhoun County. There appear to be several thousand acres of swamp forest east of Marysville extending linearly along the Apalachicola River in Liberty and Calhoun Counties. Fishing and hunting camps and logging activity are prevalent in the region, and disturbance by humans is obvious. By far the most promising area in this region is located near the confluence of the Chipola and Apalachicola to the east of Wewahitchka in Gulf and Liberty Counties. Because of the vast forested backwater areas and lack of landmarks, I was advised not to go into this area by canoe without a local guide and was unable to obtain one, thus my efforts were restricted to woods roads and drier sites, which failed to yield encouraging results.

Additional exploration is in order, but because all the forest I saw was second growth and relatively young, I decided that other sites warranted investigation before the commitment that would be needed to mount an expedition. The remoteness and promise of this region is enhanced by the presence of Apalachicola National Forest and the Apalachicola Wildlife Management Area in southwestern Liberty County. Several other smaller river swamps (e.g., the Sopchoppy River) occur within the national forest, adding to the overall quality and extent of the habitat.

Florida was certainly the heart of ivory-bill country and may yet offer the best hope for the survival of the species. But my years of following tantalizing leads and probing Florida's remaining wild areas have thus far been fruitless. Sadly, with each return visit to once prime ivory-bill habitat I find increasing evidence of human activity and fragmentation of forest habitats.

Georgia

Most Georgia records are pre-1900; however, reported Okefenokee Swamp sightings of ivory-bills have been as recent as 1948. C. N. Elliott described possible nesting of ivory-bills on Minnie's Island in the Okefenokee in 1931.[31] In his report to the National Audubon Society for the first half of 1937, Tanner noted that he had talked to E. Adams, a technician at Okefenokee, who had recently seen ivory-bills there. Tanner lamented that he had no time to return for more investigations. Pending the receipt of additional information, I had considered valid the sighting of November 30, 1948, at Gap O'Grand Prairie in the Okefenokee by Philips B. Street and F. V. Hebard.[32] Hebard also noted, "There's another recent report from the west end of the south fork of the Canal. Our Coleraine birds were last recorded late in April, 1946."[33] I discussed this record at length with Street, and he indicated that he never saw the bird, had been looking in the opposite direction when Hebard had seen it, and was surprised to find his name associated with the record. He seriously questioned the observation.

Stoddard reported seeing an ivory-billed woodpecker at a distance of about fifty yards while he was flying in a small plane over the Altamaha

Plate 1. Louis Jean Pierre Vieillot's painting of a male ivory-billed wood-pecker (his Plate 109, "*Le Pic noir, a bec blanc*"). From *Histoire naturelle des oiseaux de l'Amerique Septentrionale* (Paris: Desray, 1807). *Courtesy Albert M. Greenfield Digital Imaging Center for Collections, Academy of Natural Sciences, Philadelphia*

Plate 2. Cuban scientist Giraldo Alayón examining cerambycid beetle tunnels exposed by ivory-billed woodpecker bark scaling in northeast Cuba, March 1987. *Courtesy Jerome A. Jackson*

Plate 3. The large larvae of cerambycid beetles that ivory-bills seem to favor in both the United States and Cuba. I dug these larvae from a recently dead pine in northeastern Cuba on which an ivory-billed woodpecker had been foraging. James Tanner photographed similar larvae that ivory-bills were eating in Louisiana, and examination of stomach contents of ivory-bills collected more than a century ago also revealed the importance of these insects in the diet of ivory-bills. *Courtesy Jerome A. Jackson*

Palte 4. Cuban scientist Pedro Rosabal standing in front of one of the largest Cuban pines we found near Ojita de Agua in March 1988. This was the area where ivory-billed woodpeckers were found in 1986 and 1987; note the recent stump in the foreground. *Courtesy Jerome A. Jackson*

Plate 5. Photo of ivory-billed woodpecker tail showing the tip's unusual downward curvature. *Courtesy Jerome A. Jackson*

Plate 6. Photo showing differences in the bills of adult and young ivory-billed woodpeckers. The juvenile female *(top)* with the less acutely pointed bill when viewed from above is Yale University specimen #4633, purchased from General Biological Supply House, Chicago, Illinois, April 26, 1939; a second tag on the specimen suggests the bird was collected January 15, 1890, at Kissimmee, Florida. The specimen of an adult male (more acutely pointed bill) shown is Yale specimen #4632 collected on an unspecified date in Florida. *Courtesy Jerome A. Jackson*

Plate 7. A pair of ivory-billed woodpeckers with cerambycid beetle larva.
Courtesy Guy Coheleach

Palte 8. Ivory-billed woodpeckers featured on stamps from around the world. *Courtesy Jerome A. Jackson*

Plate 9. Advertisement showing a "rare liqueur" being revealed by a rare ivory-billed woodpecker. *Courtesy Jerome A. Jackson*

Plate 10. Color photo taken with Dufaycolor film of a male ivory-bill at its nest in the Singer Tract in northeast Louisiana, March 7, 1938. *Courtesy James T. Tanner*

River basin in 1958.[34] He also saw a pair in beetle-killed spruce pine (*Pinus glabra*) near Thomasville in what may have been the same year.[35]

My searches in Georgia focused on the Altamaha River and Okefenokee Swamp—a function of Stoddard's reports from the Altamaha region and the large area of protected swamp forest and long history of reports from the Okefenokee. However, later, in digging through Fish and Wildlife Service files, I found an intriguing letter reporting a pair of ivory-billed woodpeckers seen along the Ogeechee River in 1973. The report was submitted in 1985, only after publicity about the ivory-bill, but it had apparently never been followed up by the Fish and Wildlife Service. The author, Rev. C. Deming Gerow, and his son Jim knew South American ivory-bills (*Campephilus* spp.) and this might have been a good lead. I did not see the letter until 1991, but with high hopes, I wrote to Reverend Gerow at the address given. Luck was with me—he was still at that address. Gerow had served as a missionary in Argentina for thirty-four years and was a bird bander, amateur taxidermist, and especially interested in wood-peckers. He and his son were familiar with the appearance, vocalizations, and tapping of *Campephilus* woodpeckers as a result of their work in Argentina. They had caught and banded a relative of the ivory-bill—the cream-backed woodpecker (*Campephilus leucopogon*) in Argentina. Their sighting of an ivory-bill on the Ogeechee River about twenty-five miles west of Savannah, Georgia, on July 26, 1973, should have been taken seriously. Reverend Gerow told me that they were first attracted to the presence of the bird by its calls, which were somewhat similar to those of the cream-backed woodpecker. When they looked, they immediately recognized it by the white shield on its back. They observed the bird casually as it worked on a dead stub across the river. Having only recently returned from Argentina, they were not aware of the rare status of the species and hence did not report it at the time.

As a result of publicity concerning my searches, I received assistance from members of the Atlanta Audubon Society who canoed down the Altamaha in 1987 in search of ivory-bills—but with no success. My efforts along the Altamaha were limited to McIntosh, Glynn, Long, and

Wayne Counties between Jesup and Interstate 95 in January 1987 and June 1988. Southeast of Madray Springs the Altamaha Swamp widens, so that at US highway 82/25, which crosses the swamp between Jesup and Ludovici, the swamp is about 2.5 miles wide. It narrows again to about a mile wide near the McIntosh–Long County line, then widens again southeast of the Glynn–Wayne County line in its course to the coast. The mixed-swamp forest is heavily cutover and used by hunters and fishermen, but fingers of it extend up the many tributaries, such as Penholoway Creek southeast of Jesup. The area in McIntosh County southeast of Altamaha Park is partially protected as the Altamaha State Waterfowl Management Area. Because of its heavy use, the presence of a population of ivory-bills seems unlikely. However, because of the size of the swamp and adjoining areas of forest, it would certainly be possible for a pair to go undetected for long periods of time.

Louisiana

The best recently accepted records of ivory-billed woodpeckers anywhere are those from the Singer Tract (now Tensas River National Wildlife Refuge) just south of Tallulah. These records resulted from the work of Allen and colleagues in 1935 and the extensive work by Tanner in the late 1930s.[36] Following completion of Tanner's fieldwork in the Singer Tract in 1939, there was little documentation of the fate of the birds. However, two ivory-bills were observed there in August 1941, and Roger Tory Peterson saw two females in May 1942 and noted that a single ivory-bill was seen there six months later.[37] Pough found a lone female there in 1943, Peterson stated that one "was still in the Singer Tract as late as December 1946," and local reports suggest one may have been present as late as 1948, as discussed in the previous chapter.[38]

Others have gone to the Singer Tract since in search of ivory-bills, but most have come away only with dreams of what might have been. A few have reported success in their efforts—but no documentation to satisfy the scrutiny of the scientific community. Ornithologist Laurie Binford believes that he, Burt L. Monroe Jr., Delwyn G. Berrett, and

Keith A. Arnold may have heard an ivory-bill call on private land along the Tensas near Tallulah on March 11, 1962. He noted, "The call was typical of Ivory-bill—just like records, very full and rich, not like a jay imitation. Heard once very close and several times at a long distance. Binford and Monroe fairly well convinced."[39] Habitat in the area was second-growth hardwood forest with scattered big baldcypress and some large logs. Binford and party were actively searching for an ivory-bill in response to local, undocumented sightings.

On November 19, 1981, George Heinrich and C. Welch heard what they believed to be the calls of two ivory-billed woodpeckers in flight at the Singer Tract, but did not see the birds.[40] They returned to the area in mid-April 1982 and, after three days of searching, heard a call they believed to be an ivory-bill on April 15. Their next visit, in June 1982, resulted in no further evidence of the birds.

I made several visits to the old Singer Tract between 1987 and 1989. On early visits, I drove back roads on and near the refuge to examine the extent and nature of habitat, and hiked into areas where the forest was extensive and seemed mature enough to possibly support ivory-bills. During a visit to Tensas on May 21, 1989, I met briefly with refuge forester Dan Taberrer and law enforcement officer Barry Jordan and discussed condition and management plans for the forest. Neither considered it likely that ivory-bills still existed on the refuge.

I hiked old trails and logging roads and followed the banks of the Tensas River and the fringes of some of the many bayous. On one visit, ornithologist William E. Davis and I hiked/waded the approximately seven-mile roundtrip toward Rainey Lake from the main road leading to refuge headquarters. Much of the forest was underwater. We found the forest at the northern half of the refuge to be largely mature second growth with scattered relicts that might be suitable for ivory-billed woodpeckers. Tree species composition was essentially as reported by Tanner: dominated by southern red oak, willow oak (*Quercus phellos*), sweetgum, American elm, and sugarberry (*Celtis laevigata*).[41] Understory varied from open to dense switchcane (*Arundinaria gigantea*) thickets close to the Tensas River. We heard all southern woodpeckers except yellow-bellied sapsuckers, red-cockaded, and ivory-billed woodpeckers. The southern half of the Tensas Refuge is much more cut over, but still includes extent, and age, and species diversity such that it could be used by ivory-bills.

Aerial surveys along the Tensas River on May 20, 1989, impressed me with the extent, quality, integrity, and remoteness of the forest. In addition, there were numbers of large dead trees scattered throughout the forest. Although this forest is surrounded by agricultural land and has some three-wheeler trails going into it, most of it remains isolated to the general public, and I feel that there is a remote possibility that ivory-bills could remain in the area. The likelihood of this, however, becomes increasingly remote with the length of time that refuge personnel work in the area without any sign of the birds' presence.

One of the most intriguing—and frustrating—reports of ivory-billed woodpeckers is that presented by George Lowery to several ornithologists at the American Ornithologists' Union annual meeting held in Baton Rouge, Louisiana, in 1971. Lowery had obtained two color prints (Figure 9.1) of photographs taken of an ivory-billed woodpecker on two different trees on May 22, 1971, somewhere south of US highway 90 (contra Dennis,[42] who places them close to I-10) near or in the Atchafalaya Basin of Louisiana.[43] The photographs were greeted with scientific skepticism, and many believed they were of poorly stuffed birds hung on a tree.

Lowery was bitter over the matter and would not reveal the precise locality. He did not claim to have seen the birds. Rather, he said that a dog trainer, who was out with his dogs, had seen the birds fly across a right-of-way clearing. The dog trainer had a camera with him, so he pursued the birds and was able to get two photos of a male ivory-bill when the bird landed on a tree in front of him. The photos are now on file in the VIREO collections at the Philadelphia Academy of Natural Sciences, but had been at Louisiana State University (LSU) and could not be found when the Ivory-Billed Woodpecker Advisory Committee

Figure 9.1. Photo of an ivory-billed woodpecker taken by a man known only as the "Chief" in south Louisiana in 1971. *Courtesy Van Remsen, Louisiana State University Museum of Zoology and VIREO (Visual resources in Ornithology), Academy of Natural Sciences, Philadelphia*

met there in January 1986. In May 1989, I again visited LSU and was able to examine and photograph the photos. They were beginning to fade and discolor—perhaps because of their being kept in a high-acid paper envelope. The question remains as to whether or not the photos are of a live bird. The pose is certainly similar on the two trees, and neither the feet nor bill are showing.

Lowery made several trips to the area, but was never able to find the birds or feeding signs that would suggest their presence. On his first trip to the site, he located a fresh excavation in a baldcypress within about a hundred yards of where the photos were taken, but thought perhaps the cavity was being made by a pileated woodpecker. Tanner also saw the cavity and was convinced that it had been made by a pileated woodpecker.

Bruce Crider, one of Lowery's former students, told me that the

"Chief"—the man who took the photos—had several photos other than those given to Lowery and that some of those photos are partially out of focus, proving movement of the bird. Crider has not revealed the specific identity of the Chief nor has he obtained copies of those additional photos. At the Advisory Committee meeting on January 29, 1986, Crider indicated that the Chief was intensely "anti-Fed." Crider also indicated that the photos that Lowery had were not taken within the levee system of the Atchafalaya Basin but rather farther south. He stated that the landowner said that the birds were present as recently as the winter of 1981 to 1982 and that the only habitat change in the area was the clearing of a pipeline right-of-way. Crider stated that he had made three trips to the area where the photos were taken, two in the spring of 1981 and one in the spring of 1982, but saw no sign of the birds. The habitat, on hunting club land, he said included about 600 acres of baldcypress–tupelo gum—some virgin timber—surrounded by about twenty miles of mixed swamp forest.[44]

Another Atchafalaya ivory-bill report was a bird seen by Robert Bean, director of the Louisville (Kentucky) Zoo on November 11, 1974. This bird flew at car-top level across I-10 about twenty miles west of Baton Rouge. Robert Hamilton saw what he thought was an ivory-billed woodpecker in the same area in 1973.[45]

In some of his searches in the Atchafalaya, Crider has been accompanied by ornithologists John Moroney, Hamilton, and others. In March 1978, I accompanied Crider and several graduate students for two days into the Atchafalaya near the junction of Bayou Pigeon and Grand River to follow up on a possible ivory-bill that Crider heard. The habitat there was dominated by large sugarberries, several oak species, and sweetgum. Many of the trees of these species exceeded thirty-one inches in diameter, and there were numerous dead trees and an abundance of other woodpecker species in the area. We found nothing, but Crider reported at the first Advisory Committee meeting that he briefly saw an ivory-bill from a helicopter in the area a few days after our 1978 visit.

The Atchafalaya Swamp is the largest true swamp remaining within North America.[46] It is located in south-central Louisiana, formed by a distributary of the Mississippi River that flows through the basin between the natural levee on the west banks of the Mississippi and Bayou La Fourche and the natural levees of the Bayou Teche. Extensive levee systems, built following a tremendous flood in 1927, now confine much of the Atchafalaya River to a single channel, preventing it from spreading into the surrounding forests.[47]

The effects of this taming of the Atchafalaya have been negative for the ivory-billed woodpecker on both sides of the levees. Between the levees, floodwaters are now higher than ever and extensive siltation occurs. As a result, the forests between the levees are now primarily black willows (*Salix nigra*). Outside of the levees, the protection from floods has allowed the clearing of forest land for soybean (*Glycine max*) and sugarcane (*Saccharum* sp.) fields, gas and oil exploration and refining, and growing human communities.

In recent decades, canals constructed by the oil industry have not only destroyed swamp habitat and altered flow patterns but have also provided avenues for human intrusion into the most remote areas. As we flew transects back and forth over the Atchafalaya, we saw boats of sport fishermen everywhere. The National Wetlands Research Center of the US Fish and Wildlife Service provided me with a map indicating habitats within a large portion of the Atchafalaya area and east of the current "basin" known as the Lake Verret, Louisiana Project Area. Within this area, they tabulated 45,000+ acres of dry bottomland hardwoods, 52,800+ acres of transitional bottomland hardwoods, and 179,200+ acres of bald-cypress-tupelo. These areas in addition to thousands of acres of sweet-gum-oak and baldcypress forest along the eastern and southwestern edges of the basin provide perhaps the best as well as most extensive habitat remaining anywhere for the ivory-billed woodpecker.

I made several visits to the Atchafalaya between 1986 and 1989. My efforts included canoeing in the Bayou Pigeon area, driving back roads leading into the basin between Lafayette and Morgan City, and Morgan City and Plaquemine, hiking logging roads and through mature forest areas, and aerial reconnaissance over the entire basin from Interstate 10 south to near Morgan City. I was unable to rent a boat in the area, and motor problems precluded use of my own boat for getting deep into the basin. I made contacts with a number of local biologists, seeking information about habitat and possible sightings of ivory-bills, without getting any positive feelings about the prospects of finding ivory-bills in the area. All were familiar with pileated woodpeckers but all also assumed that the ivory-bill is extinct.

On May 19, 1989, in a Cessna 172, Davis and I flew transects over the basin at an altitude of 500 to 800 feet. At that height we felt we might be able to identify prime habitat and perhaps even an ivory-bill in flight. Within the levee, essentially the whole western two thirds of the basin was dominated by black willows—the result of the flooding regime. Prime sweetgum-oak forest was most prevalent at the northeastern end of the basin, with the Bayou Pigeon area being among the better forest habitats. Baldcypress was common in the northwestern and southwestern parts of the basin. Outside of the levee, there is some good sweetgum-oak habitat south of Lake Fausse Point. I hiked through some of this area in March 1989, but found that there were few large trees and that human developments were so close that I was never out of hearing range of vehicles and boats.

To the extreme northeast, just south of I-10, is a promising area that is posted by Tenneco Oil Company; we were unable to obtain permission to enter, however. North of I-10 and along the east shore of the Atchafalaya River we found what appeared to be suitable habitat within the Sherburne Wildlife Management Area (11,700+ acres) and Atchafalaya National Wildlife Refuge (15,200+ acres).

As with areas lower in the basin, access roads for oil wells were useful in allowing us to see the habitat but also indicative of the level of disturbance. Like the canals, these roads changed water-flow patterns, fragmented forest, and opened up the habitat for human intrusion. Sweetgum and several oak species were prominent, along with boxelder (*Acer negundo*), elm (*Ulmus* sp.), sugarberry, sycamore (*Platanus occidentalis*), red mulberry (*Morus rubra*), and cottonwood.

Flooding prevented us from getting far enough from access roads to take meaningful measurements, but the forest was generally young with well-distributed older trees—mostly of noncommercial species. We played the ivory-bill tape at several locations without response. Other woodpecker species were common. Because of the flooding, I could not adequately evaluate the potential for ivory-bills in this area; association with the larger Atchafalaya Basin to the south increases the potential for ivory-bills in this area, and it provides yet another link in an important chain of protected bottomland forest habitats between the Atchafalaya, the Singer Tract, and areas along the Mississippi and Yazoo Rivers to the northeast.

Along the border of Louisiana and Mississippi are bottomlands of the lower Pearl River. A large area of the lower Pearl River bottomland is included as buffer lands for the Mississippi Test Facility, where rocket engines are tested in Hancock County, Mississippi. Adjacent to this site is the Honey Island Swamp that straddles the Mississippi-Louisiana border, extending for fifteen to twenty miles along the river and varying in width from about four to seven miles. The lower Pearl River includes a wilderness of approximately 185,000 acres, with about 92,000 acres of that under protection as either part of the Pearl River Wildlife Management Area or the Bogue Chitto National Wildlife Refuge.[48]

I have visited this area many times in the past thirty years, including five days during the spring of 1979 when I was doing environmental surveys at the Mississippi Test Facility, specifically looking for evidence of ivory-billed woodpeckers, Bachman's warbler, and red-cockaded woodpeckers. Efforts included floating down the Pearl River and aerial surveys of forest habitats using a Cessna 172.

The Mississippi Test Facility is generally covered by young forest—some of which is in pine plantation. Along the east side of the Pearl River system there are scattered large hardwoods, but these are few. In contrast, the Honey Island Swamp area has many large trees scattered throughout—including baldcypress of eighty inches in diameter. These tower above the rest of the forest and are easily viewed from the I-10 bridges that pass through the swamp.

I was puzzled by the lack of ivory-billed woodpecker records from the Pearl River Basin in Tanner's tabulation, but at the time was not familiar with the history of the area. I found the answer in S. G. Thigpen's history of the Pearl River.[49] During the nineteenth century, Pearlington, just north of the Mississippi Test Facility, was one of the most important lumber centers in the United States. Lumber from the Poitevent and Favre Lumber Company at Pearlington was used to build the L & N Railroad between Louisville, Kentucky, and Montgomery, Alabama. It was also lumber from this mill that was used to build jetties at the mouth of the Mississippi, and by the beginning of the twentieth century one of the owners was known as the "lumber king" of the South as a result of worldwide sales of Pearl River lumber. In 1895 one of the Pearlington mills was billed as the largest lumber mill in the world. In short, it seems likely that there are no ivory-bill records from this region because the establishment and operation of these mills resulted in destruction of the bird's habitat before anyone was particularly interested in documenting their presence.

By the time Tanner was researching ivory-bills, the Pearl River forests were completely cut over. Just now the protected areas of the Honey Island Swamp, Pearl River Wildlife Management Area, and Bogue Chitto National Wildlife Refuge are reaching a maturity that might once again be capable of supporting ivory-billed woodpeckers.

On April 1, 1999, Louisiana State University graduate student David Kulivan was turkey hunting in the Pearl River Swamp of southeast Louisiana when he observed at close hand what he believed were a pair of ivory-billed woodpeckers. Kulivan had previously seen pileated woodpeckers and his descriptions fit. However, considerable effort to find the birds turned up no evidence of them. In 2002, as a result of the Kulivan sighting, an organized search, sponsored by the Carl Zeiss Optical Company, generated considerable excitement in the birding community. On January 27, 2002, the searchers heard and recorded what they immediately thought was the double rap of an ivory-billed woodpecker.[50] I was contacted and asked for my opinion of the sounds. To me they were clearly not the rapping of an ivory-bill; there was no loud *"BAM"* followed immediately by a lesser *"bam."* To me they sounded like gunshots—clear, evenly spaced, and identically sounding: *"bam bam bam bam."* I responded with my opinion that they were gunshots and was told, "You'll be sorry you said that." And immediately I was, but it *was* my opinion. I was concerned about my interpretation because I was listening to a recording and the searchers had heard the sounds out in the Pearl River Swamp. The recording, made by Martjan Lammertink, also clearly captured the sounds on tape. Independent acoustic analysis of the sounds at the Cornell Laboratory of Ornithology and at Harvard demonstrated that they were gunshots.[51]

Efforts by the Cornell Laboratory of Ornithology in conjunction with the Zeiss search provided great hope for evidence that ivory-billed woodpeckers might be in the Pearl River Swamp.[52] In late January 2002, John Fitzpatrick and colleagues installed twelve acoustic recording units in good habitat within the Honey Island Swamp area of the lower Pearl River. These sound-activated recorders were left in place through mid-March, then retrieved, and the sounds recorded compared against the acoustic signatures of the ivory-bill calls recorded by Allen. None of the sounds recorded was that of an ivory-bill. Furthermore, the technology provided additional evidence that the mysterious loud *"bams"* recorded by Lammertink and heard by other members of the ivory-bill search team were gunshots.

In my several visits to the area, I have never seen sign of ivory-bills, although I would like to believe they might be there. One scenario that I have considered is that ivory-bills displaced from the Singer Tract moved to forests along the Mississippi—such as perhaps those near Vicksburg where we had a possible ivory-bill respond to a recording (see below). Then, with cutting in that area, they could have moved farther south to the Pearl River area. We put speculation aside, for now we have a thin ray of hope. J. Rosen and J. J. Williams provided good general accounts of Kulivan's and other recent observations.[53]

Mississippi

Although there are no well-documented reports of ivory-billed wood-peckers in Mississippi during this century, M. G. Vaiden of Rosedale told James Bond of the Philadelphia Academy of Natural Sciences that he knew of six pairs in hardwoods nine miles south of Rosedale until the forest was cut during World War II.

A published report of ivory-bills in central Mississippi approximately thirty miles north of Meridian in a "very isolated region" as recently as 1953 is tantalizing, indicating the observer knew the differences in appearance and call between the ivory-billed and pileated woodpeckers, but the report lacks details.[54]

My ivory-bill searches took me to four areas of Mississippi: Black Creek and the Pascagoula Swamp along the lower Pascagoula River in southeast Mississippi; the lower Yazoo River, Delta National Forest, and other areas near the Mississippi River from Memphis south to the Louisiana border; the Tombigbee River and Tibbee Creek area near West Point and the Tombigbee and Noxubee rivers south of Columbus, in east-central Mississippi; and the lower Pearl River Swamp in southwest Mississippi and southeast Louisiana.

In 1978, ornithologists Ronald Sauey and Charles Luthin visited southeast Mississippi to float Black Creek, a meandering blackwater stream that flows through DeSoto National Forest, eventually joining the Pascagoula River in extensive swamp forest. On the second day of their float they heard what Sauey later wrote "sounded every bit like the historic Ivory-bill recording of Allen and Kellogg." The following is from Sauey's letter of February 2, 1978, to me:

> On our second day of boating on the creek (floating without motor to be as quiet as possible) we found an amazing congregation of mixed species—Brewer's [blackbird, *Euphagus cyanocephalus*], Rusty [blackbird, *Euphagus carolinus*], Redwings [red-winged blackbird], Orange-crowneds [warbler, *Vermivora celata*], Yellow-rumpeds [warbler, *Dendroica coronata*], etc., etc., etc., and dozens of woodpeckers. We were

probably moored . . . for about an hour when a couple of Pileateds flew in . . . and started up a ruckus. Shortly after, we heard a very loud series of tappings from farther down the river and then a number of distinct musical calls, given repeatedly on the same pitch and reminding us both of a nuthatch [*Sitta* sp.], only louder and not as nasal. The calls stopped, and then were repeated again, only closer this time to us. The call sounded even less like a nuthatch the second time, being fuller and more resonant, and we both looked at each other in disbelief—was it an Ivory-bill? . . . we never saw the creature making the call.

Sauey's report, combined with reports of others, the vastness of the Pascagoula Swamp, and the history of ivory-billed woodpecker specimens collected in the area in the late 1800s all suggest that the swamp forests of southeast Mississippi hold promise for ivory-bills. Straddling the boundary between George and Jackson Counties in southeast Mississippi, the Pascagoula Swamp is a bottomland hardwood swamp of about 42,000 acres. In 1976, approximately 32,000 acres of the swamp gained protection when Mississippi purchased it.[55] Ivory-billed woodpeckers were reported from the Pascagoula River Swamp as recently as 1921, but until Sauey's 1978 letter, I knew of no more recent reports.[56] Then on March 2, 1982, I received a call from Judy Toups, a good friend and competent coastal birder. The previous week, a birding friend of hers, Mary Morris of Biloxi, reported seeing two ivory-billed woodpeckers on her property on the west side of the Pascagoula River, south of the Wade–Vancleave Road and east of the Old River Road north of Vancleave. Morris first heard a "honking" noise and thought that geese were flying over. When she looked up to determine the source of the honking she saw two large woodpeckers on a pine near her. At first she thought they were pileated woodpeckers, but then she realized that the back edge of the wings was white. She also noticed that each bird had a crest and the crest of each pointed toward the other bird. When the birds finally flew, she saw that the whole trailing edge of the wing was white. Toups went to the site with Morris on the following day, but did not see nor hear the birds.

She played the Cornell recording, but got no response. On March 4 and 5, my wife and I went to the site with Toups. We found no evidence of ivory-bills. The area was cutover pinelands near the Pascagoula Hardwood Tract, but there were a lot of human activities in the area and a good number of houseboats and cottages.

I made other visits to the area in December 1986, and April, June, and July 1988. These included floating down the Pascagoula from near Boneyard Lake in George County, to the Wade–VanCleave Road in Jackson County. I also drove, hiked, and waded along logging roads in southern George County and the northern half of Jackson County with no success.

The description provided by Morris was perfect and also unique. I have never heard the call of an ivory-bill likened to the honking of geese, but it is a reasonable comparison. The forests of the Pascagoula Swamp are cutover but include many relicts. D. G. Schueler mentioned a baldcypress nine feet in diameter.[57] I found none that large, but recorded several oaks, sweetgums, sycamores, and sugarberries three feet or more in diameter. I got no responses to playback of ivory-bill calls on any of my visits—but it is a big swamp.

Well to the northwest of the Pascagoula River, in west-central Mississippi, flows the Yazoo River. Four independent but inconclusive reports of ivory-billed woodpeckers along the lower Yazoo River and adjacent Mississippi River bottoms came to me as a result of initial publicity for my searches. My attention had been drawn to the area because it is only about thirty miles from the Singer Tract.

Between October 1986 and May 1989, I visited many areas along the Mississippi and Yazoo Rivers. Some offered potential for ivory-bills, but I quickly focused on an area of about 12,300 acres of bottomland hardwood forest between the Yazoo and the Mississippi. The area is essentially an island that is isolated by the two rivers on the east, west, and south, and on the north by an oxbow formed at the old mouth of the Yazoo. Although this forest is accessible as a result of a ferry that crosses the Yazoo River, it is private land managed by hunt clubs and is closed to the general public. I had difficulty gaining access to it as a result of flooding in the spring and hunting seasons in the fall.

My first visits to the area were by boat, when I explored the flooded backwaters near the old mouth of the Yazoo. I made later visits on foot and flew over the area to determine the contiguity and character of forest habitat and to assess the distribution and abundance of dead trees.

On March 29, 1987, my graduate student, Malcolm Hodges, and I were just south of the old mouth of the Yazoo in mature bottomland hardwoods. Dominant trees in the area were sweetgum, sugarberry, water oak (*Quercus nigra*), southern red oak, hickory, honey locust, and boxelder. We had begun at dawn and were following the procedure of walking fifteen minutes, stopping, playing the tape for forty-five seconds, and then listening for three minutes. We were impressed with the habitat. At about 11:00 a.m. we came to a low area where there were numerous large (bigger than forty inches in diameter) trees, some of them recently dead.

I commented to Malcolm that based on Tanner's habitat descriptions, this was by far the best habitat I had seen. I played the tape—and waited. I heard no response. Then after about a minute, Malcolm was visibly excited, whispering loudly "There it is! Listen, it's coming this way." I heard nothing. At least another minute passed, all the time Malcolm indicating that the bird was still calling. Then I heard it. A bird was giving precisely the call that was on the tape and it was coming toward us from the southeast.

The bird continued to call and to come closer so that it became clearly audible. We judged it to have stopped a little over 150 yards from us. We waited, hoping that it would continue to approach, but it did not. We played brief segments of the call, but it remained in the distance, occasionally calling.

Finally we decided that we had to get close enough to see and photograph it, so we began moving forward—and it was gone. We last heard the bird at 11:28. Later, in another part of the forest, we had a blue jay (*Cyanocitta cristata*) respond to the recording, giving similar calls, but always ending its call with a typical blue jay note.

Although we have no documentation, Malcolm and I both believe that the bird that responded to our tape was an ivory-billed woodpecker. The bases for our opinion are the combined facts that the bird flew in from a great distance; it was extremely wary and would not let us approach closer than about 150 feet; it called incessantly for several minutes, giving the call from the tape with invariable cadence and pitch; we never heard a blue jay call at that site; and the habitat seemed excellent for ivory-bills (mature hardwoods, numbers of large recently dead trees, high numbers of other woodpeckers). We searched the remainder of the day, resumed playing the tape an hour later, but had no further responses. At the time, I had only a playback machine with me.

On describing these events to a friend, it was suggested that having the blue jays imitate the ivory-bill call might in itself be indicative of the presence of ivory-bills. I disagree. A few weeks after this incident I was standing in a friend's backyard in northern New Jersey when a blue jay began to give this very call—far outside of the range of any ivory-bill and in the absence of any taped stimulus. Again, the New Jersey blue jay soon lapsed into more typical blue jay calls.

On August 11, 1988, ornithologists Davis and Fred Sibley were near the mouth of the Yazoo River, within a quarter of a mile of where we heard the bird respond to the tape, when they heard a similar call. They did not have recording equipment or an ivory-bill tape to play, but both felt that the single *"toots"* that they heard were identical to those on the Cornell tape. This time the bird was heard for only a few seconds. They never saw it.

Subsequent visits to the area produced no additional evidence of ivory-bills. I found no scaling of bark that might have been the work of a foraging ivory-bill. There were many other woodpeckers there—seven species recorded—and there were many pileated/ivory-billed-sized cavities in trees scattered through the forest.

Unfortunately, on March 20, 1989, while visiting this area, we met with one of the caretakers of the land who indicated that there were plans to sell more than a million dollars worth of timber from the tract during the next year. That forest is now gone.

On August 18, 1988, Charles T. Bryson, a botanist with the US Department of Agriculture, wrote to me with details of a possible ivory-billed woodpecker sighting north of the Yazoo River and northeast of Greenwood at the edge of an old slough on May 8, 1988:

> The bird was observed from a vantage point . . . at the edge of the water on the north side of the gravel road. . . . The observation was made from a distance of about 150–200 feet for a few seconds as the woodpecker flew from the west to the east across the swamp at about $^2/_3$ the height of the Tupelo gum trees. The bird's size was similar to that of a Pileated Woodpecker but, because of the shadows and the afternoon lighting, only black-and-white markings were apparent. . . . To my astonishment, it did not sound like a Pileated Woodpecker. The call was a rhythmic "toot, toot, toot," which lasted the duration of the sight contact and for a few seconds after the bird had disappeared into the Tupelo gum canopy.

I had visited this area in November 1986, and did so again in October 1988. This is the headwaters area of the Yazoo, and the area of Bryson's sighting was near bottomlands of the Yalobusha River. The area between the Tallahatchie and the Yalobusha Rivers includes extensive forested bottomlands and several oxbow lakes. Most of the area is fragmented by agricultural lands and I could find no habitat of sufficient extent or forest age that seemed worthy of investigation—in spite of the clarity of Bryson's description. I played the recording at several locations, but without response. It seems likely that if the bird was an ivory-bill, it must have been wandering through the area.

South Carolina

The most recent South Carolina sightings that have been confirmed by multiple observers began with observations of George M. Melamphy in 1934.[58] Melamphy found ivory-bills in the Wadmacon Island area of the Santee River Swamp in Georgetown County. The species' presence was confirmed by observation of a bird there in May 1935 by Alexander

Sprunt Jr., and Lester L. Walsh. Several ornithologists saw the birds in the area over the next two years.[59]

More recent reports have not been confirmed. The 1961 Report of the Committee on Bird Protection of the American Ornithologists' Union reported that a "reliable observer" had recently reported one ivory-bill in South Carolina.[60] At the first Ivory-Billed Woodpecker Advisory Committee meeting, reports were mentioned from McBee (Chesterfield County) and from the Congaree Swamp. Tammy Dabbs of Columbia, South Carolina, claimed that ivory-bills nested in the area as late as 1969 to 1970. She learned of the birds from a dentist who had reported that he knew a Native American who had ivory-bills nesting on his land. Supposedly a contact at the South Carolina Game and Fish brought ridicule and the dentist was "turned off." Dabbs followed up on the report and she and two other people went to the Black River, where they heard two ivory-bills and saw silhouettes of two on trees. She was told to contact Paul Sykes of the US Fish and Wildlife Service. Sykes searched habitat in the area from September 12 to 15 and on September 20, 1970, and reported some beautiful swamp forest but found no evidence of ivory-bills.[61] An ivory-bill was supposedly seen in the area as recently as 1981.

In 1971, Robert Manns made national news with a report of an ivory-bill having responded to a tape recording in a swamp along the Santee River near Columbia.[62] I discussed this report with Manns several years ago and it seems likely to me that the bird responding might have been a wood duck.

My visits to South Carolina included time spent along the Black, Cooper, Great PeeDee, Santee, Waccamaw, and Savannah Rivers, including Charleston, Berkeley, and Georgetown Counties. One ivory-bill report from this area came in via Lester Short from Baker Doyle, a trustee of the American Museum of Natural History. His report was of a bird seen on the lower Waccamaw River, but I was not able to obtain details of the sighting. I visited the area following the report and found mostly cutover forest, pine plantations, and numerous fishing–hunting camps.

Most of the areas visited in association with the first five rivers were in Charleston, Berkeley, and Georgetown Counties.

Francis Marion National Forest borders the South Santee River, and Wambaw Creek flows through the forest to empty into the Santee. These combined areas provided several thousand acres of bottomland swamp with many adjacent stands of mature pine (until Hurricane Hugo destroyed much of the pine in 1988). One potential ivory-bill area in this part of South Carolina was designated as the Wambaw Creek Wilderness Area in 1980.[63] In 1974, South Carolina ornithologist E. B. Chamberlain, who knew the area well, suggested that the Francis Marion might still be home to ivory-bills.[64] From what I saw of the wilderness area, it could be.

I visited the wilderness area and adjacent forest lands in May 1989. In the wilderness area it was nearly impossible to hike cross-country because of intermittent standing water and a tangle of fallen trees. For about the lower seven miles of Wambaw Creek, the swamp is a little more than half a mile wide. Farther upstream it widened to more than a mile wide. The most impressive trees in the area are the loblolly pines (*Pinus taeda*) that often exceeded three feet in diameter, but in back waters and low areas along the creek there were equally large sweetgum and oak trees. I heard or saw all seven species of woodpecker that were possible in the area for this time of year—everything other than the ivory-bill. The ivory-bill potential was enhanced by an abundance of large dead trees. Although the wilderness area is relatively small (about 1,900 acres), it is surrounded by several thousand acres of national forest land and private timberlands.

One area of importance is Wadmacon Island, just to the north of the junction of Wambaw Creek and the Santee River. This island bounded by Wadmacon Creek and the Santee River is about two miles wide and six miles long. Immediately to the northwest is Santee Swamp, and more to the north is bottomland area associated with Cedar Creek. Much of the latter area has been cleared and planted in pine plantations. Although this area has a past history of timber exploitation and parts of the swamp were cleared for rice farming in the 1800s, the second growth is now mature and promising. Foremost among factors that argue against the presence of ivory-bills in these areas is the extensive use of the area as research sites by competent field biologists who would readily recognize an ivory-bill. Most of the ornithologists are, however, working with red-cockaded woodpeckers in the upland areas on the Francis Marion and on the Hobcaw. Other negative factors are ubiquitous fishermen in any area that is accessible by boat or road, and the Wambaw Creek area is advertised as a place for canoeing buffs.[65] Certainly if ivory-bills are not already in the area and were found somewhere else, and a population became available for introduction, this area would be a prospect that would have to be considered.

Along South Carolina's western border with Georgia, the Savannah River Swamp is a relatively narrow swamp (generally less than a mile wide) that extends along the east bank of the Savannah River from just south of Augusta, Georgia. My visits to the area of Savannah River were numerous, including more than twenty-five days in mature swamp forest between Augusta and Garnett between 1974 and 1987.

My efforts in the South Carolina counties of Aiken, Barnwell, and Allendale included aerial surveys of that portion of the Savannah River swamp that is now a part of the US Department of Energy's Savannah River Plant and areas south to Groton Plantation in Hampton County. I spent several days canoeing and wading in the Savannah River Swamp on the Savannah River Plant specifically to survey the bird life there.

This swamp includes a few enormous baldcypress (eighty-plus inches diameter) and many two- to three-feet-diameter sweetgums and oaks. Much of this swamp is protected as a part of the Savannah River Plant nuclear facility. Other areas that I visited and flew over included the Silver Bluff Plantation of the National Audubon Society and areas just downstream in Allendale and Hampton Counties that are protected as parts of private game preserves. One of these was the Groton Plantation, Tanner visited in 1937.[66] Tanner noted that there were about 7,000 acres of uncut bottomland swamp there at that time. This area had been maintained as a hunting preserve and, although some timber was sold, the forests were managed by selective cutting. There was extensive old growth present, both in the uplands and in swamp forest, but I could not determine the extent of the latter. I believe that there was suitable habitat for ivory-billed woodpeckers in the area, but I felt that the extensive ecological work done by personnel associated with the Savannah River Ecology Laboratory would have turned up some evidence of them sometime during the previous thirty years.

The only potentially positive sign of ivory-bills in the area was a large cavity that ornithologist Robert McFarlane and I examined on the Savannah River Plant in the mid-1970s. It was old and seemed too large for a pileated cavity, appearing quite like some of the cavities in old photographs of ivory-billed woodpecker nest sites. But there was no conclusive evidence of the maker's identity. The nest and roost cavities of ivory-billed and pileated woodpeckers are similar in size and somewhat variable in shape, although often elongated, perhaps to accommodate the crest that each of these birds has. Although identifying a cavity as being that of an ivory-bill would be difficult, the imprint of bill marks made during excavation might provide strong evidence, because the ivory-bill has a much more chisel-tipped bill, whereas the pileated's bill is more pointed. Old cavities sometimes have a few feathers left in them, and those too might be used to identify the former occupant.

Texas

The ivory-billed woodpecker was not reported in Texas after 1906 until sightings in the 1930s reported by Bessie M. Reid and Corrie H. Hooks in counties east of the Trinity River.[67] These were never confirmed, although Eastman mentioned a twenty-four-page manuscript by Reid, a newspaper columnist, detailing observations of ivory-bills in the Big Thicket from 1901 to 1956.[68] The manuscript was sent to Robert Porter Allen of the National Audubon Society, who considered the observations reliable. Tanner spent parts of four days with Reid, and he found that her reports of ivory-bills came entirely from others, and those he was able to speak with did not convince him they had seen the birds.[69]

The most recent reports from Texas were of observations made in the 1960s in the Big Thicket area between the Trinity and Neches Rivers by Eastman, Dennis, and others.[70] In 1962, Eastman suggested that there were two pairs and one extra female remaining in a near virgin forest area in the Big Thicket.[71] Dennis reported hearing and recording an ivory-bill there on February 25, 1968. The recording was later analyzed at the Bioacoustic Laboratory at the Florida Museum of Natural History by John William Hardy, who concluded that it was either a blue jay or an ivory-billed woodpecker.[72] Then he threw "scientific conservatism to the wind" and stated "my ear leans toward the ivory-bill."

In response to Dennis's reports from the area, Tanner and Sykes spent a week in January 1968 with Dennis in the Big Thicket, searching areas in the Neches and Angelina River bottoms and along a section of the Trinity River.[73] They searched on foot, by car, boat, and plane and Sykes and Tanner concluded that there were no ivory-bills there. George Miksch Sutton also investigated these ivory-bill reports and indicated that he felt Dennis was "overly optimistic."[74] Tanner also noted that Ernest McDaniel, a native of the Big Thicket and former president of the Texas Ornithological Society, had searched for ivory-bills in east Texas for ten years with no success.[75]

Secondhand reports of the Big Thicket ivory-bills seem to have used considerable poetic license. For example, British writer Philip Street, in 1971, incorrectly noted that "In 1967 . . . several pairs were discovered by an ornithologist working for the Bureau of Wildlife of eastern Texas."[76]

A record by Neal Wright, an unemployed woodsman, is intriguing and yet inconclusive. Wright was said to have found a pair at a nest in the Big Thicket (no locality given) in March 1967 and had photographed the female in the nest entrance.[77] Reynard saw the photo and said that it was fuzzy but definitely of a *Campephilus* woodpecker. The bizarre ending of the story is that the area was "agent-oranged" and cut shortly after the discovery. No further details or reasons for the actions are known.

Hal Owens, a Texas lawyer and amateur birder, searched repeatedly for ivory-bills in the Big Thicket and had what he referred to as an "extremely likely sighting" in 1969.[78] He had used distribution of posters offering a cash reward for verified ivory-bill sightings (Figure 9.2), aerial surveys, and amplified recordings with no positive results.

Bob Mounsey and a team from the University of the Wilderness reported two sightings of ivory-billed woodpeckers near Steinhagen Reservoir in the Big Thicket area on May 21 and 22, 1976.[79] Dennis believed these records were valid.[80]

Reynard provided a 1969 record that is particularly intriguing—a sound recording that he says is of the double rap of an ivory-bill that he recorded in the Big Thicket. The recording is included on a phonograph record of "Bird Songs in Cuba."[81] Lacking recordings of the ivory-billed woodpecker from Cuba, Reynard substituted a portion of the recordings made by Peter Paul Kellogg and Arthur Allen in Louisiana in 1935—plus a brief and poor-quality recording of the "double-knock" of an ivory-billed woodpecker, which Reynard recorded in the Big Thicket in 1969.

I provided Tanner with a copy of the recording, to which he responded,

> I hear on this tape 64 seconds of recordings of Ivory-bills made in 1935 in Louisiana, followed by about 9 seconds of fuzzy sound with the sequence "b-bam, cawing of crows, an unidentified wail"; this sequence is immediately repeated. I do not believe that the "b-bam" is the double rap of an Ivory-bill. That sound is more deliberate, more time between the two blows, and the second rap is weaker than the first.[82]

My searches in Texas were limited to the Big Thicket area. Fortunately, in 1974, Congress set aside 84,500 acres of east Texas as the Big Thicket National Preserve. Certainly one of the factors weighing heavily on the mandate to set aside this area was the report of ivory-billed woodpeckers by Dennis and others in the area in the late 1960s. Ornithologists, however, remained skeptical of the discovery, and Dennis himself later said "it would be almost a miracle if any are still holding on in its diminished woodlands."[83]

Figure 9.2. "Wanted Poster" used by Hal Owens in his efforts to find ivory-billed woodpeckers in the Big Thicket. *Courtesy Hal Owens*

The Ivory-Billed Woodpecker Advisory Committee agreed that the Big Thicket was a low priority area, but I spent part of October 1987 in areas of Polk, Tyler, and Hardin Counties anyway to see what remained of forest habitats. This superficial visit revealed only cutover hardwoods and pine plantations. With the efforts of the University of the Wilderness groups and others following up on Dennis's reports, and with the reward offered by Hal Owens, if ivory-bills are still in the Big Thicket, it would seem that we would have more evidence than is at hand.

CHAPTER 10

Cuba

"*No problema,*" the forest guard repeatedly told us in a dry,
    nonreassuring voice.
"*Problema,*" I kept saying to myself.

*I*t seemed days had passed, but it was only one morning (back in
April 1987, and well before dawn) that we had left the city of
Guantanamo. I was in Cuba for the first time, having been asked by
Lester Short from the American Museum of Natural History in New
York to join him, Jennifer Horne from the Kenyan National Museum
in Nairobi, Christoph Imboden from the International Council for Bird
Preservation (ICBP) in London, and Montserrat Carbonell from ICBP
in Buenos Aires. We were there to evaluate conservation efforts in Cuba,
but I was there, in truth, because the ivory-billed woodpecker had just
been discovered in the mountains of eastern Cuba. Cubans had found
them in 1986 and Short, Horne, and George Reynard had verified the
sighting with their own observations.

The trip to the Ojito de Agua area in the mountains west of Baracoa
was made in a Russian Niva, a vehicle about the equivalent of an
American Jeep. We started on highways that narrowed to country roads,

Figure 10.1. Crossing the Rio Toa in March 1987 in northeast Cuba, where
ivory-billed woodpeckers were found in 1986. *Courtesy Jerome A. Jackson*

turned to gravel, then dirt, then barely passable lanes. When we reached
the Rio Toa it was still early morning. Our driver quickly decided that
the water was too deep for us to drive across. A radio call brought an
enormous tractor; a rope was produced, and our vehicle was towed
across, bobbing like a cork in the river (Figure 10.1). On the other side,
the eroded and abandoned logging road had three-feet-deep ruts in
places. After many stops, digging, filling, and pushing, we decided to
continue our trek upward on foot. It was mid-day when we arrived at
Ojito de Agua—"eye of the water"—a spring high in the mountains.
This was the place where ivory-bills had been found.

Excitement was high. But I knew the trip was a quick one. One day
is all we could spend in this land of ivory-bills, and most of that time
would be spent getting there and back. Ours was more of a diplomatic,
fact-finding mission than a field expedition—and that night we were
scheduled to have dinner with the governor of Guantanamo Province,
the province where part of the Ojito de Aqua area is located. So there I
was, so close to ivory-bills, yet knowing that my short stay would probably prevent me from seeing them.

As we reached the ivory-bill habitats the tension grew. I spotted a sign. The pines clearly showed evidence of ivory-bill feeding. There were dead snags with huge slabs of bark that had been knocked off to reveal the one-inch-diameter tunnels of cerambycid beetle larvae. The larvae had bored just under the bark, and ivory-bills had stripped the bark away to reach them (Plate 2). Everywhere I looked I hoped for one glimpse of an ivory-bill. After only about four hours in ivory-bill country, we were told it was time to leave. "Dinner and the governor are waiting," I wrote in my notes.

As we started down the mountain on foot to the Niva, the sunny day turned overcast, and then it began to rain. Soon there was mud everywhere. Slippery mud. Far more slippery than I had remembered mud being. In hopes of getting close to an ivory-bill, I had brought a good deal of camera equipment and I was paying a dear price for my decision. I really was not prepared for the downpour. By the time we reached the Rio Toa, the waters were at flood stage. There was no way our Niva could cross this time.

"*No problema,*" the optimistic forest guard said again.

Four miles upstream he said there was a dam. Then two and a half miles beyond that there was a village. But the Niva could not go that way. He told us we would need to walk to and over the dam, then to the village and ask to borrow a vehicle. We started hiking up the river. It rained endlessly. Then it got dark. With no flashlights we made our way over eroded limestone the Cubans call *diente de perro*—tooth of the dog, because it is so sharp.

My leather boots were now saturated with water. I felt ridiculously loaded down with telephoto lenses, cameras, and binoculars that I was trying to keep dry. The rain and mud coated my glasses and I could not focus on even the small bit of light that was getting through the dark night. I lost count of how many times I slipped and fell, but I am sure, looking back, that the scene resembled a silent comedy. At the time, however, I was in pain from twisting my left ankle. My water-soaked boots were cut by the sharp limestone to the extent that they began to fall apart.

Then a stick came through the loosened sole of my boot and jammed into my toe. The sole of the boot started doubling back on itself, tripping me as I walked. Then the sole simply fell off, leaving the upper part of my left boot still laced to my ankle. Now I was walking barefoot over *diente de perro*! I lagged behind, tortured by the walk, my exposed foot so sore that I wanted to stop, but I could not. Dinner with the governor seemed as far into the future as the peeled bark of the ivory-bills seemed in the past. I was a robot, forcing myself to put one foot in front of the other to keep up with my companions. Finally we stopped.

"*Aqui!*" The forest guard said.

"*Donde?*" I asked. We had reached the dam, but I could see nothing in the darkness. I could only hear the rush of the river over the steady rain. I waited for him to show us where and how we were to cross.

"*No problema,*" the forest guard said.

"*Problema,*" I said. "We can't cross here!"

The forest guard picked up a stick and poked into the water. It was still raining, but it seemed to be letting up. The dam, or something solid, lurked a foot or so beneath the surface of the water. We linked arms and he led us through the swirling water and darkness. Excepting our guard, we had no idea how wide the top of the dam was nor how wide the river was. After a tense walk we reached the other side, climbed the bank, and continued toward the village. Up and down in the darkness, we crossed another stream—chest deep, but narrow. I was physically exhausted and my left foot was in incredible pain. Then I remembered the hundred dollars.

Before leaving Mississippi, I had neatly hidden a $100 bill under the liner of one boot. Which boot? Had I lost my $100? On that first trip to Cuba I was more than a bit apprehensive. Americans were not supposed to go to Cuba. If I had $100 hidden away, it might be useful in getting me out of any jams I might encounter. The thought of the $100 kept me thinking and moving. Had it been in my left boot?

Finally we reached a road and I collapsed. I was clearly in the worst shape of anyone and was somewhat embarrassed by it. But pain has a way of diminishing embarrassment. I could tell that Short and the others were concerned. The forest guard left us and headed for the village. After about thirty minutes he returned in a school bus that he had somehow acquired from the village around midnight on an incredibly rainy night.

We arrived back in Guantanamo and I could hardly wait to get to my hotel room. But we did not go to the hotel. We went to the governor's mansion instead. Dinner and the governor were still waiting. One shoe off, one shoe on, covered with mud and still soaked, we had dinner with the governor of Guantanamo Province at 2:00 in the morning.

Back in my hotel room I pulled off my right boot. There, beneath the liner, was my $100—neatly folded—a white piece of paper that, when I opened it up, had only one square of printed green left on it. When I got back to the United States, my bank readily exchanged it for a new bill, but now I wish I had kept it as a souvenir of my first Cuban search for ivory-bills.

Early History

Our knowledge of the ivory-billed woodpecker in Cuba begins with Johannes Gundlach, who was born in Marburg, Germany, in 1811 and emigrated from Germany to Cuba in 1839. Gundlach spent most of the rest of his life studying Cuban and Puerto Rican natural history. Although he published much in his native Germany, he also contributed to Cuban and North American journals. He adopted Cuba as his home and was known in Cuba as Juan Gundlach.

Gundlach's association with the ivory-bill began within a few years of his arrival in Cuba. While collecting birds he discovered a trio of ivory-bills, one of which had an incredibly long, curved bill. No doubt knocked out of alignment by an injury of some sort, the bill could not wear properly, and when Gundlach found the bird in 1843, the bill was seventeen inches long. How could such a bird eat? Gundlach had to know. He followed the trio for three days, noting that the two normal birds occasionally fed the long-billed bird but that the unusual bird was also able to poke into large arboreal termite nests and extract termites for itself. Such

Figure 10.2. Specimen of the female Cuban ivory-billed woodpecker with an anomalous seventeen-inch bill. *Courtesy Jerome A. Jackson*

seemingly altruistic behavior has been noted occasionally in birds, but rather than being altruistic, this behavior may instead reflect a long fledgling dependency in the species. His curiosity satisfied, Gundlach shot all three and prepared them as taxidermy mounts for display. Charles B. Cory examined this specimen at Gundlach's private museum at Ingenio Fermina, Cuba, and illustrated it along with a line drawing in *The Auk* in 1886.[1] Cory illustrated the bird as an adult male. I examined the specimen, now in the collections of the Cuban Museum of Natural History in Havana, and it is an adult female (Figure 10.2).

Gundlach had a long history with the ivory-bill, providing specimens to many museums and assisting others in obtaining specimens. Richard C. Taylor presented a pair of Gundlach's ivory-bill specimens to the Philadelphia Academy of Natural Sciences in about 1846. Gundlach presented a pair to the Zoologishe Museum in Berlin in December 1849. Another pair of Gundlach's ivory-bills was presented to the Musée d'Histoire Naturelle in Paris. Charles Wright presented a pair to the US National Museum in 1861 via Gundlach as well. In 1884 to 1885 and again in 1887 to 1888, Gundlach was sent by Dr. Carlos de la Torre, curator of the Museum of the Institute of Havana, and by Dr. Fernando Reynoso, director, to eastern Cuba to secure "additional examples of rare local species." At the end of the nineteenth century Cuban forests were rapidly disappearing—and so was the Cuban ivory-bill.

The Cuban ivory-billed woodpecker was described as a new species, *Campephilus bairdii*, by John Cassin of the Philadelphia Academy of Natural Sciences in 1863, writing that it appeared to be "one of the singular insular species which have become well known to naturalists."[2] Cassin noted that in all respects except size and a few other minor characteristics, the Cuban ivory-bill was quite like the North American ivory-bill. He named it after his friend, Spencer Fullerton Baird, secretary of the Smithsonian Institution, whom he says suggested to him that it might be a new species. However, the new species designation of the Cuban ivory-billed woodpecker was not generally accepted, and Baird, T. M. Brewer, and Robert Ridgway, in their 1874 classic *A History of North American Birds*, described the Cuban bird as a subspecies of the North American ivory-bill (*Campephilus principalis bairdi*).[3] They suggested that it could be distinguished from the North American birds by two plumage characteristics. The white neck stripe of the Cuban ivory-bill is said to reach the bill, but it does not extend that far in the North American form. Also, on the adult male Cuban birds, the scarlet feathers of the crest are said to be longer than the black crest feathers, whereas they are the same length as the black feathers in North American birds.

Following Gundlach's discovery and collection of the first ivory-bills from Cuba in the 1840s, knowledge of the birds grew slowly. Ramon de la Sagra reported seeing one in October 1850, but locality details were not mentioned.[4] Understanding of the former distribution of the species in Cuba is that it was widespread in extensive forested areas throughout the island, including both upland and lowland and pine and hardwood forests (Figure 10.3).

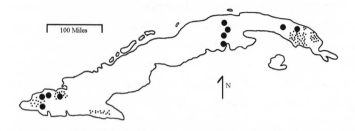

Figure 10.3. Map of known localities of the Cuban ivory-billed woodpecker.

Gundlach reported their decline even before they were said to be in trouble in North America.[5] By the end of the nineteenth century they seem to have disappeared from all areas but the mountains in eastern Cuba: Sierra de Moa, Sierra del Cristal, Sierra de Nipe, and possibly Sierra Maestra.[6] In 1923, Witmer Stone, in reviewing Thomas Barbour's *Birds of Cuba*, commented on the extinction of some Cuban birds and added "the Cuban Ivory-billed Woodpecker . . . has apparently but recently reached the end of the journey."[7]

With exploitation of the Cuban forests, ivory-bill populations were fragmented, and one by one they disappeared. The deforestation in Cuba was in part to provide wood for a growing population and industrial economy but mainly to provide land for growing sugar cane. Following the Spanish-American War, the United States gave Cuba "favored nation" status and guaranteed to buy all the sugar Cuba could produce. US sugar interests quickly moved in and forests were cleared for sugar cane even in areas unsuitable for growing cane. With clearing of the forests came collectors seeking the birds and *campesinos* (rural people) who hunted the ivory-bill as food.[8] The last known specimen from Cuba was collected by C. T. Ramsden in 1907 at San Luis de la Cabezada, Guantanamo Province.[9]

Conservation efforts for the Cuban ivory-bill began at the start of the twentieth century.[10] On January 18, 1909, Article 19 of Cuban government Decree Number 67 noted that all Cuban woodpeckers, including *El Carpintero Real*, the ivory-bill, were "useful" birds and therefore could not be killed or captured at any time of year. A 65,460-acre refuge called Parque Nacional Refuge located in the estate known variously as *El Cristal, Realongo del Cristal,* or *Sierra del Cristal* in Oriente Province included a population of ivory-bills at least until 1910. Unfortunately, the laws seem to have had little effect on the conservation of the birds.

By 1923, Barbour described the Cuban ivory-bill as "virtually extinct," suggesting that there might be a few pairs still living in the mountain pine forests of Mayarí.[11] By 1938, as the fight to save the last ivory-bills in the United States was being lost, some believed that the Cuban ivory-bill was extinct.[12] James Bond, noted ornithologist and author of *Birds of the West Indies*, held out some hope, believing that the Cuban ivory-bill still existed but was confined to the Sierra del Cristal in Oriente Province, where it had last been seen by Swedish botanist Erik Ekman in about 1920.[13] Although Barbour, again in 1943, echoed Abelardo Moreno's sentiment that the Cuban ivory-bill was extinct, he acknowledged some possibility that a few pairs may have survived in the pine forests of Mayarí, where Ramsden had collected a pair in the early 1900s.[14]

Mid-Twentieth Century

In March 1948, C. Russell Mason, executive director of the Massachusetts Audubon Society, got word of the possible existence of ivory-bills in the mountains of eastern Cuba. He discussed this with two of his employees, Davis Crompton and John Dennis, and noting their excitement and determination to find the ivory-bill, gave them a month off to go to Cuba to search for them. Their story has a somewhat fairy-tale twist. They had heard what they believed were ivory-bills, traipsed after them, but were unsuccessful. Although separated, each had stopped to rest and had fallen asleep, only to be awakened by an ivory-bill hammering at the bark of a tree that was near them both. On April 17, 1948, after days of searching hardwood forests, Dennis and Crompton found the birds in a remote, burned-over, cutover pine forest.[15] They located three birds, including a pair incubating at a nest thirty feet up in a rotted pine stub. A photo of a male they observed at the nest clearly shows the open nature of the habitat. Logging in the region was extensive at the time, and the area where the ivory-bills were nesting had been logged about seven years earlier. Although Dennis and Crompton heard of another nest nearby, they were not able to visit it.

In 1956, Bond reported the Cuban ivory-bill to be "apparently confined to northeastern Oriente Province (from the pinares of Mayarí eastward)."[16] He also noted that it was said to have inhabited the mountains of Pinar del Rio Province as late as the early twentieth century. Bond was probably citing information provided by George Lamb who, in 1956, found twelve or thirteen ivory-bills near La Melba, in northeastern Cuba.[17]

Lamb had received a fellowship from the International Committee for Bird Preservation (Pan American Section) for a year's study of rare species in the Caribbean. First on their list was the ivory-bill. Lamb and his wife spent about five months in early 1956 searching for the birds near Moa in northeastern Cuba. In a letter to Whitney Eastman, Bayard Read, a professional wildlife photographer and Lamb's father-in-law, noted, "It was only by camping out in the mountains for weeks at a time that they saw any birds."[18] Read was planning to go to Cuba to photograph the birds if they found a nest, but they did not. Indeed, the only time they ever saw the birds was before daylight or at dusk. Nonetheless, Lamb provided basic information on roost cavities, behavior, distribution, and habitats of the birds.

Following the Cuban Revolution in 1959, scientists from the United States and many other countries were unable to study or even search for the ivory-bill in Cuba, and the status of the species was unknown for several years. Cuban scientists continued searches for ivory-bills, but information about the searches is scanty. Orlando Garrido saw a female ivory-bill about 5 miles north/northwest of Cupeyal and 12.4 miles south of Moa along the road to Moa, high in the mountains in early February 1968.[19] He saw the bird as it perched on a dead tree thirty feet away. Kai Curry-Lindahl suggested that by 1956, only twelve or thirteen individuals remained in eastern Cuba but that in 1969 there were thirteen pairs.[20] No source for this population estimate was given, and without more information I am inclined to think this is an error of a later transcription from Lamb's "thirteen birds" to "thirteen pairs." As with other authors, Curry-Lindahl suggested loss of old-growth forest as the cause for the species decline.

Renewed Hopes

In 1981, Eugene Morton of the Smithsonian Institution reported that as of November 1980 one pair of ivory-bills remained in the area of Cuba where Lamb had found the birds in 1956.[21] By November 1981, birders were offered guided tours to the area, noting "this may be the last opportunity to see this bird before it becomes extinct."[22]

Some reported sightings of ivory-bills in Cuba are in doubt, such as a 1982 report from the western Cuba province of Pinar del Rio.[23] A report of the species on the Isle of Pines (*Isla de Juventud*) is clearly in error.[24] But the ivory-bill did survive in Cuba past the middle of the twentieth century and possibly still exists. The last refuge of the species in Cuba seems to have been the pine forests of the Sierra de Moa in Oriente and eastern Holguin Provinces in northeast Cuba, and possibly areas in the Sierra Maestra in southeast Cuba.

In 1985, Short, Reynard, and Giraldo Alayón visited the Cupeyal Reserve, just west of the area where Lamb observed the birds.[25] They found fresh feeding signs and heard of a sighting that had been made in December 1984.[26] In October 1985 and March 1986, Alayón and Alberto Estrada made additional searches in the area, following Lamb's itinerary. They found evidence that the birds were near Ojito de Agua, not far from the Cupeyal Reserve. Then on March 13, 1986, Estrada got a glimpse of an ivory-bill near Ojito de Agua, and on March 16, Alayón saw a female ivory-bill being chased by two Cuban crows.[27]

Within a month, Short, Horne, and Reynard returned to Ojito de Agua. They were joined by Estrada and Alayón, and they found at least one and possibly two ivory-bills. Their efforts to photograph the birds or record their calls or double rap were unsuccessful.[28]

To the ornithological and birding world, the rediscovery of ivory-bills in Cuba was electrifying. The news received national and international coverage. Short's exclamation "I saw it!" captured in the magazine *International Wildlife,* revealed the simple human excitement of finding what has long been sought, even by a man who has seen and studied more species of woodpeckers than any other.[29] Estrada and Garcia reacted to it more profoundly: *"Es realidad, no un sueño"*—it's real, not a dream![30]

Into Ivory-Bill Country

Back in the United States, with my feet recovered and new boots purchased, I began efforts to return to Cuba. In early 1988 I was asked to serve as scientist on a National Geographic Expedition to the mountains of eastern Cuba in search of the ivory-bill. I left Starkville, Mississippi, on February 21, 1988, and in Miami met Robert Hernandez, a *National Geographic* editor, Ted Parker, a highly respected ornithologist, and Bates Littlehales, a *National Geographic* photographer. We arrived in Havana at 4:30 a.m. on the 22nd. Later that day we attended a meeting with representatives from the Cuban Academy of Sciences at the Casa de Scientificos, a sort of hostel for visiting scientists. Politics and planning were first. A man named Gomez, the vice president of the Academy of Sciences, formally greeted us. We had an obligatory drink of rum; Gomez left; then we settled down to business. The briefing was useful, with the Cubans giving us a topographical map of the Ojito de Agua/Jaguani area—actually a section cut from a larger topographical map, and we learned later that the particular map was a "classified" military document. They also gave us copies of the agreement with *National Geographic,* a detailed plan for the expedition, and a review of the history of ivory-billed woodpecker observations in Cuba.

I outlined a search strategy of starting early, before dawn each day if possible, moving quietly in ones or twos, wearing no red or white clothes because of the potential for negative reactions from the birds, and moving for fifteen minutes, stopping to listen and look for fifteen minutes, then moving on again. The Cubans seemed genuinely impressed with our collective expertise and seemed to have done a great deal of work in preparing for the expedition. I had, over the course of the previous year, developed a good working relationship with some of the Cuban scientists, and it was beginning to pay dividends.

The next morning we set out by plane for Santiago de Cuba. At the Santiago airport we were greeted by a group of mariachis, who serenaded us with "Guantanamera" and other local music. From the Santiago airport, we went to the local Casa de Cientificos to await our equipment and to rest. Later we were taken on a tour of Santiago de Cuba and the vicinity—to San Juan Hill, then to a new resort about eleven miles east–southeast—the Hotel Buccanero. There were many tourists at the Hotel Buccanero, and all that we spoke with were Canadian. A plateau came nearly to the water's edge, and the resort was perched at the base of the fifty-foot cliffs that reached nearly to the Caribbean. We walked up a narrow canyon where I found and photographed the nest of the larger of Cuba's two native hummingbirds, the Cuban emerald (*Chlorstilbon ricordii*). We also saw a number of other native birds as well as North American migrants: Cuban tody (*Todus multicolor*), Oriente warbler (*Teretistris fornsi*), Cuban pewee (*Contopus caribaeus*), Cuban bullfinch (*Melopyrrha nigra*), black-throated blue warbler (*Dendroica caerulescens*), magnolia warbler (*Dendroica magnolia*), American redstart (*Setophaga ruticilla*), Cape May warbler (*Dendroica tigrina*), northern parula (*Parula americana*), black-and-white warbler (*Mniotilta varia*), cave swallow (*Petrochelidon fulva*), belted kingfisher (*Ceryle alcyon*), turkey vulture (*Cathartes aura*), and Cuban gnatcatcher (*Polioptila lembeyei*).

We left the hotel and traveled a couple of miles northeast to the *Valle de Prehistoria* and found an enormous hilly savanna populated by dozens of lifesize, realistically modeled concrete dinosaurs. Apparently this was a major amusement park in the making. Our next stop was a small house where Fidel Castro had organized his 1953 attack on the Moncada barracks. It is now a little museum—of which we got a personal tour. The entire area we visited has been declared a national park—*Bacanao*—that is rapidly being developed as a major tourist attraction. The park extends from the sea well inland to the mountains, and includes several beaches, campgrounds, and many historical sites. The landscape is dry—much like southern California—except for an abundance of native palm trees.

The next day we traveled to Guantanamo, where I stayed at the Hotel Guantanamo along with the three *National Geographic* staff members; our Cuban colleagues stayed at a local government guest house. It was a slow process getting to the woodpecker habitat, but we had to move on Cuban terms. I felt I was gaining a better understanding of the human dimension that was necessary to consider in any conservation effort. That evening, after dinner, we spoke with our host, Gladstone Oliva, about Cuban life. I took down the following notes:

> minimum wage in Cuba is 80 pesos/month
>
> 2.6 percent of the population gets minimum wage
>
> 10 percent of the wage of one of the wage earners in a family is the rent on their home
>
> after twenty years of paying rent, they own the home
>
> Gladstone gets 575 pesos/month
>
> professors at the top level make 400 pesos/month

Politics was the dominant theme of discussions. Oliva strongly defended and happily described the Cuban system. A lot of comparisons were made to Brazil, as well as to the United States.

Rainy weather in the mountains raised the level of the Rio Toa such that we could not get into the ivory-bill area the following day. Even without the weather delay, it seemed the Cubans were still making various arrangements for our stay in the mountains. We were left to ourselves much of the day, but in late afternoon our hosts took us on a field trip in search of the Zapata sparrow (*Torreornis inexpectata*). We drove to a narrow band of scrub near the Caribbean, twenty-five miles east of Guantanamo. It was an incredibly hot and uncomfortable ride in a Romanian jeep. When we stopped, we searched the scrub in foothills on the inland side of the coast highway. To the north lay cliffs composed of the familiar *diente de perro*. An abundance of cacti and agave were just beginning to bloom. We quickly found the sparrow, although sightings were fleeting and I never got a chance to take a decent photograph. Our bird list from the trip also included dozens of Cuban grassquits (*Tiaris canora*); several Cuban gnatcatchers; a light-phase Cuban kestrel (*Falco sparverius sparveroides*); turkey vultures; Cape May, northern parula, Oriente, palm (*Dendroica palmarum*), and prairie warblers (*Dendroica discolor*); red-necked pigeons (*Columba squamosa*); common ground-doves (*Columbina passerina*); mourning doves (*Zenaida macroura*); Cuban emeralds; and Cuban vireos (*Vireo gundlachii*). I spent most of my time photographing or my list might have been longer.

Back at the hotel that night, Mayito—an environmental officer for Guantanamo Province—came to see us and let us know that he would be going with us. He had also accompanied us into the area in 1987. Mayito brought a doctor to meet us and discuss our health needs. A Cuban photographer who was supposed to work with the *National Geographic* crew also arrived. I learned the doctor was also to accompany us. The Cuban photographer certainly did not seem to be the "outdoor type"—nor did he seem to know a great deal about the camera equipment we had, although he was nice. The expedition seemed to grow and grow and to be motionless.

Although we were all tired and anxious about the next day's journey to the ivory-bill area, Oliva insisted on giving us a tour of the city of Guantanamo that evening at 10 p.m. Nothing could be done to avoid the tour, although every attempt was certainly made. So we piled into the vehicles and drove to the square at the city center. Oliva said it was a custom to walk around the park at night—and so we did. The square was clean and neat, with permanent benches lining the periphery and facing both the street and the center of the park. There was a lighted "band shell" with what was supposed to be a pool in front of it. The pool was dry, but there were two four-foot diameter concrete water lilies in its center. There was also a neatly painted Catholic Church in the square, which Oliva was quick to note still "worked."

Next to the church was a statue of José Marti, the Cuban poet and journalist who had become the symbol of Cuba's struggle for independence from Spain. It was in a well-lighted shrine with plastic flowers and a little fence around it. There were probably three hundred people in the park—most in small groups just sitting on the benches. There was no loud music or noise, no noticeable drinking, no graffiti on the walls. It was quite a contrast to late-night teenage gatherings at outdoor spots in the United States. Surrounding the square were a number of stores with well-lighted windows displaying clothes, shoes, and fabrics as well as a theater and a social club built by the Spanish in the nineteenth century. The architecture of the square was beautiful, and I was soon glad Oliva had insisted we come.

After leaving the square, Oliva took us to a monument that was just across the street from the Hotel Guantanamo. At night the monument was impressive, slightly reminiscent of Stonehenge. There were several concrete pillars, some with concrete statues and others with concrete faces—all with names and slogans relating to the two Cuban revolutions. Early the next morning I walked back to the monument so that I could take some photographs. Concrete is a poor man's marble, and, viewed by day, it looked awful in places. But the Cubans were extremely proud of it. Oliva had to visit the Institute of Geography Office in Guantanamo to make arrangements for our trip, and I tagged along. I was quickly ushered into a display area and shown the Russian space capsule in which a Cuban astronaut rode. The display also included numerous photographs of the launch and recovery, plus the space suit worn by the Cuban.

By mid-morning of February 25, we had completed final preparations and finally left Guantanamo for Los Rusos—literally, "the Russians." We were told it had been the site of the camp of a Russian team of geologists. Los Rusos saddles a ridge in the heart of the Ojito de Agua area in which ivory-billed woodpeckers were seen by Short and his party in 1986, and it is at the junction of two major trails and near the headwaters of the Yarey River. The trip to Los Rusos was a pleasant, uneventful one. Beautiful trees were in bloom in the foothills all along the way, lighting our path with a brilliant orange. We stopped in the village of Palenque to pick up a forest guard and make additional arrangements. Our next stop was a school, which Oliva informed us was named Old Mother Number Five. There we picked up food and supplies. There were many students milling around, but no sign of classes in progress. The students were most eager for us to photograph them, and we obliged. From there it was only a short distance to where we turned off the main road onto the trail leading to Los Rusos. My first impression was that, unlike the previous year, the road was dry and easy to negotiate. The waters of the Rio Toa were low and we had no trouble fording it. Once we got into the higher elevations things changed. The road was muddier, but we only got stuck once. At one point we found a pair of Cuban parrots (*Amazona leucocephala*) perched on a dead stub next to the road. *National Geographic* photographer Littlehales excitedly sprang into action. He used the jeep as a blind and twisted his body like a pretzel to get the shots he wanted. I was in the back seat on the left side of the vehicle but managed to get a couple of shots before they flew off.

At 1:00 p.m. we finally arrived at Los Rusos and began setting up camp. Our tents were similar to ones that I have used elsewhere—nylon, airy, free-standing, and comfortable. The Cubans set up their olive-drab canvas two-person wall-tents in the pines at the east edge of the clearing. We set up a six-person tent in the center of the clearing to hold supplies and provide sleeping quarters for the cook and driver.

Oliva headed back to Guantanamo shortly after we arrived. As soon as we got the tents up and our gear inside, the rain began. It rained off and on all afternoon and most of the night. The tents did not leak, but I was frustrated. Here we were, eager to get into the field, and weather kept us in the tents. Between bouts of rain, I walked west to the spot from which ivory-bills had been seen and pointed out the beetle work and old ivory-bill work on dead pines to the others. My immediate impression was that little had changed since I had been at the site in the year before—there did not seem to be any fresh signs of ivory-bills.

I noted on February 25 that there were numerous plants in bloom and many more about to come into bloom. This was supposed to be the dry season, but from the looks of the flowers, the ferns, and the sky, I wondered what the wet season was like. For supper our first night at Los Rusos the cook prepared some of the dehydrated trail food that we had brought with us. He could not read the English directions and just improvised. It did not turn out too badly, however. Unfortunately the Cubans seemed enamored with our fare and it seemed as if we might quickly exhaust the trail food. Boxes of chocolate-covered granola bars and packages of M&M candy provided by the *National Geographic* were almost gone the first night. By morning there were granola bar and candy wrappers strewn everywhere.

On the first night, Littlehales became quite ill, seeming to have flu-like symptoms. He did not get up for breakfast—and I should not have. The cook, experimenting with the unfamiliar ingredients, served us a terrible, pasty, lumpy concoction made by filling a cup nearly full with powdered milk, adding hot water, then adding one tablespoon of Nestle's Quik. After the gourmet breakfast, we set out as two groups to look for ivory-bills.

At first we traveled together, heading east down the ridge along the logging road, but parted when the road forked. One group headed south and my group of three went north. Shortly after noon we returned to camp to find Bates seriously ill. The jeep had left, heading to the little village of Riito to retrieve a doctor—the doctor we had met in Guantanamo had not accompanied us. In the midst of illness, the air was filled with the smell of lunch—canned Russian fish that seemed to me the smelliest, oiliest looking stuff I could imagine; it reminded me of cat food. The cook had emptied several cans into his big cook pan and just stirred it up over the fire. I forced a bit down and tossed the rest into the woods. I fixed Bates a cup of dehydrated vegetable soup and he drank the broth.

At 2:30 p.m. the jeep arrived with the doctor, a twenty-five-year-old, nice-looking, slender man who was friendly and possessed an air of competence. Pedro Rosabal, one of the Cuban scientists accompanying us, told me that all doctors in Cuba must spend three years working in the countryside or with Cuban troops in Angola before going on to specialize. Needless to say, many opt for the countryside. Dr. Hernandez wore a gray casual suit—muddy black shoes, open white shirt, and bright blue socks. He spoke only Spanish. I helped him at the tent with Bates, whose temperature was then 101.8°F. The doctor took Bates's blood pressure and temperature and gave him two more aspirin tablets. After leaving the tent, the doctor conferred with Hiram Gonzales, the Cuban scientist leading the trip, to go back to the clinic with the doctor. Bates was determined to stick it out, but agreed to go back in the morning if he was not better. Before the doctor left, the Cubans took him to the food tent and filled his medical bag with granola bars and candy. Two days later the fever broke, but Bates was still quite ill. He hung in there for days, dedicated to the need to photograph the ivory-bills, but finally, on March 2, it was clear even to him that he needed to be taken back to the United States for treatment. In the end *National Geographic* sent a new photographer, and Bates ended up staying in a hospital for six weeks with complications from pneumonia.

Our deep concerns about Bates's health did not keep us from searching for ivory-bills. Each day we walked along the ridges, each unfortunately with its own logging road, to assess the forests below. We quickly learned that the understory was extremely dense at lower elevations, particularly near the streams. In many areas there was a bamboolike grass that was so dense and tall that it was nearly impenetrable. To leave the logging roads meant losing the ability to cover much ground, and it also meant poor visibility. Thus, our first days were spent on the ridge roads searching for openings in the valleys below. For days we would sit silently at vantage points along the ridge in hopes of getting a glimpse of an ivory-bill passing below. On my first day of observing, a rainbow arched over some tall dead pines on a slope about three

Figure 10.4. A large dead pine, shredded and on the ground, found near Ojito de Agua in northeast Cuba when I hurried in the direction of a call that sounded like that of an ivory-bill. No other woodpeckers large enough to do this kind of work exist in Cuba. *Courtesy Jerome A. Jackson*

hundred yards below me. There were clear signs of bark scaling by ivory-bills, and I kept my tripod-mounted camera focused on those trees. This was my best chance, I thought.

In general the ridges are dominated by Cuban pines (*Pinus cubensis*), the oldest of which appeared to be thirty to forty years old. Even the oldest seemed too young to be of interest to ivory-billed woodpeckers. There were few dead trees, but the area of our camp included an abundance of dead pines, some of which were about fifteen inches in diameter, perhaps of suitable size for ivory-bill use. These were the trees that showed clear evidence of ivory-bill work, and this was one of the main areas where sightings had been made. I felt guilty that our camp seemed to sit in the best habitat for the birds, but the choice of sites had been made for us by our Cuban colleagues, and we had no option to move. If ivory-bills are at all wary, then our presence and camp activities would almost certainly have forced them to leave.

On February 29, Rosabal and I hiked southwest from camp, crossed the Yarey River, and climbed to the main road. From there we followed the road to the south, leaving it at the south edge of a clearcut. We then made our way down the slope, heading east, back to the Yarey River. Once there we waded back upstream to camp. It was impossible to merely follow the stream because of the density of the vegetation. We hiked from dawn to dark, impressed by both the diversity and density of the vegetation. The largest tree that we encountered that day was a dead pine that was fifteen inches diameter. One of our Cuban drivers told me that as recently as 1984 he worked as a truck driver removing two-foot diameter pines from the area. Loggers had seen the ivory-bills during their operations, but by 1986, when they were "officially documented," most of the large trees were gone. We found only two large pines. The first was at the edge of the bottom. It had been left standing in a log-assembly area near the Jaguani River. I measured it at two feet in diameter. The stump of an equally big pine lay just in front of it, testament to recent cutting in the area (Plate 4).

The second big pine was across the Jaguani and up the mountain slope to the northeast, along an old mining trail; it was thirty-two inches in diameter. The virgin forest that others had seen before was gone. It had been cut over, although a few old trees were scattered along the slope, standing as sentinels watching the young trees and trying to recapture a lost past.

It seemed to me that the bottom areas—some quite isolated and extensive—might be the best remaining habitat for the ivory-bill. I saw hardwood trees in some bottom forests that were more than two feet in diameter. But that was an impossible area to traverse; we could not penetrate the dense understory vegetation, nor could we have seen over it if we had fought our way through. Even beginning my travels at 3 a.m., I could make little progress before the day ended. Some of the Cubans were reluctant to accompany us because the grasses were covered with ticks. Not wanting us to be split up from them, and not wishing to accompany us, our Cuban hosts would not allow us to camp and push on from the bottoms, so we were left to do what we could in a day trip. Nor could we conduct aerial surveys, but I still think that might have been the best approach.

One day we came on piles of pine poles along a road. Although the area has been declared a national preserve for the ivory-bill, local villagers were still cutting trees. Our hosts were visibly upset. We examined the cut ends and decided they were not recent but had likely been cut within the past year and left to dry. It did not seem like the effort was of a large scale and, kept to a minimum, such "poaching" is unlikely to do much harm. Still, larger operations could begin and quickly wipe out any hope the ivory-bills have.

Encountering the Will-o-the-Wisp

My first possible encounter with an ivory-bill was at 10:15 a.m. on March 1. Rosabal and I were heading north along the road to Farallones de Moa when we came to a large dead pine lying on the ground to the right of the road. The tree was in shreds, as if broken up by a large woodpecker (Figure 10.4). As I photographed it and examined it for bill impressions (which I did not find), a bird gave a couple of loud nuthatchlike notes from just over the ridge to the east of us, about seventy yards away. The calls were definitely "ivory-bill" in nature. We walked quietly and slowly in the direction of the notes, but neither saw what had made them nor heard the bird again. On our return later in the day a yellow-shafted flicker, a poor excavator, was calling from the same site. But I knew that neither the notes I heard nor the splintering of the log had been made by a flicker.

On the next day Alayón wanted to show me what he believed was fresh ivory-bill work that he had found at Ojito de Agua. This was an area where others had reported seeing an ivory-bill. I wrote notes of the trip, recording the following:

> There's no question in my mind—the Cerambycid beetles are doing the work that everyone keeps calling ivory-bill work. I took several photos and measurements of some holes. They are very regular, sausage-shaped, with very smooth sides that go straight in. Most are about half an inch wide. One I measured was 4.7 inches long and 2.8 inches deep. No way

an ivory-bill could dig such a deep, perfectly-shaped hole. It would have to chisel in at an angle.

The ivory-bills had been there and totally removed bark from these trees to reveal the large beetle tunnels. Piles of bark slabs at the base of each tree were testament to their efforts, but the neat tunnels were the work of insects, not the bill of an ivory-bill.

My next encounter with a possible ivory-billed woodpecker came at 9:04 on the morning of March 3. Rosabal and I had been at our overlook watching the forest about three hundred feet below us since 7:45 a.m. The weather was unstable, clear one minute, overcast and foggy the next. There were also strong winds from the east most of the time, with periods of calm. It was during a period of sunny calm that we heard two nasal ivory-bill-like "*nyenk*" calls from the forest below. They stopped and we saw no bird. We stayed at our posts for two and a half more hours but detected no additional evidence of ivory-bills. Later that day, at about 3:15 p.m., we headed east down the main trail and took the southeast fork. We stopped at a good vantage point to see the forest on the south slope of the big mountain to the north. As we were discussing prospects of getting up the mountain, Alayón came running up and said he had just heard an ivory-bill along the trail near Cayo Probado. We took off at a run, getting separated as our different running speeds strung us out. All of a sudden I realized that I did not know where Cayo Probado was. I stopped and waited for the others. It was near dusk and, as my group tired, we realized we were hopelessly unsure of where to go. We headed to camp. A few minutes later those that had made it to Cayo Probado, the fast ones that could keep up, reported hearing an ivory-bill several times from about 200 yards away on the same ridge as our camp. They did not see the bird, however.

The next day, my luck changed as Rosabal and I watched from the ridge where I had photographed the rainbow.[31] My field notes tell the story:

9:32 a.m. It all happened very fast. I didn't have my camera in hand. [It was on the tripod, focused on a dead pine three hundred feet below.] But it wouldn't have made any difference. I was standing up and had just taken my binoculars down from scanning the trees below when a bird appeared right in front of me—no more than 10 yards away—headed downhill (north, not west) with wings folded back in a dive—as woodpeckers do at the top of their undulating flight. Vegetation gave me a total field of view of the bird for about 10 yards, and in that entire distance it didn't flap its wings or extend them. As it went by all I really saw was a large bird—seemingly smaller by a third (?) than the Cuban Crow—but with folded wings size would be deceiving. It was mostly black, but there was definitely white too. I didn't see the bill or crest—my total viewing time was probably about 3 seconds—but nothing else is here like it. Wow!

My next encounter was on the afternoon of March 4 at the same location. Rosabal and I took up positions at the overlook at about 3:15 p.m.

The light was bad, but I just had a feeling that the bird might come back. At 16:12 there was some slow, deliberate tapping to the south in the valley below. I was immediately up to look for the bird, but then the tapping became rapid—and I didn't think Ivory-bills would sound like that.

The Pewee was there. An occasional Cuban Solitaire *Myadestes elisabeth* called. A flock of Oriente Warblers came by. The Tody called; Trogan [*Priotelus temnurus*]; a flicker. The last of the Turkey Vultures was gone from the sky by about 16:30.

17:26. There it was! An Ivory-bill calling from behind us [to the east]—probably a couple of hundred yards or more behind us, but clear distinct toots—not at all like the Cuban Crow calls I was afraid others might have been hearing. About 5 toots were given, singly, at about 2–3-second intervals as the bird was flying uphill toward the main ridge. Then they stopped. Rosabal heard them too.

Last night Ted had heard the bird at 17:10, just over the same ridge—looks like a pattern emerging. Only one bird heard or seen. Ivory-bills supposedly forage together, but maybe there is only one—or maybe they are incubating!

On the morning of March 5, three of the Cubans reported hearing an ivory-bill between 5:30 and 6 a.m. Four days later, on March 9, I visited the mixed forest just south of the Jaguani and below the area known as Cayo Probado. As roost time approached we decided to wait at the bottom of the hill to listen and watch for ivory-bills going to roost. At 5:06 p.m. I heard four to five nasal toots that I felt certain were made by an ivory-bill. They came from west-northwest of our position at quite a distance, near Cayo Probado. I had a tape recorder running at the time, but the calls were too faint to pick up.

Abruptly the Cubans called our expedition to an end. They were simply tired and ready to return home. On March 16, we left for Havana and then on to the United States. We had not gotten the photos or recordings we had hoped for, but I left knowing that ivory-bills still haunted the forests of Cuba. I was also sure they were not common there and faced an uncertain future. Still, there was hope.

Habitat Considerations

Understanding of present habitat conditions and potential factors affecting habitat recovery in the mountains of eastern Cuba are essential to any conservation efforts. The ivory-billed woodpecker has obviously existed in diverse habitats, the common threads being old-growth forest and extensive, unbroken forest. I would add to these obvious characteristics the importance of open forest.[32] Beyond these, however, there were also habitat differences, especially tree species composition, but also likely prey-base composition. For example, termite nests were noticeably missing from the pine forests that we visited—quite different from areas near the Zapata Swamp, a lowland area in south-central Cuba near the "Bay of Pigs," where ivory-bills were once known. We found a few termite nests in the mixed forests along the rivers, adding additional support to the notion that the ivory-bills might be there. Gundlach had observed ivory-bills feeding at termite nests.[33] The Cerambycid beetle larvae are sometimes 2.5 to 3.5 inches long and, in the pines, seem to be the food that ivory-bills prefer.

In addition to the younger age of current forests, the pine forests that clothe the mountain slopes in Cuba differ in a significant way from the prehistoric forests: humans have been and are deliberately excluding fire. The network of roads and trails alone might stop fires, but as soon as one is detected, there is an effort to put it out. The result is a dense understory and, because of the pines' intolerance of shade, there is an absence of seedling pines.

In my opinion the understory in the Cuban forests is detrimental to any remaining ivory-bills. There is no way that a bird as large as an ivory-bill could fly through it. All of the ivory-bill work we found was in edge situations where the dead trees were in an area open enough to allow flight access. Furthermore, the moist environment that is conducive to existence of the cerambycid beetle larvae is only close to the ground. Higher in the dead trees the sun bakes the wood dry—an inhospitable environment for most wood-boring insects. If the remaining live pines were as large as the dead stubs, they would provide a moister microclimate, and as a result there would be more food for the ivory-bill. As it is, the few big dead trees tower above the living ones and are like so many dry toothpicks—fine as perches, not good for cavity excavation or foraging.

I feel strongly that management of the area for ivory-bills must include allowing the forest to mature and introduction of prescribed, controlled fire to open up the understory. Such fire will kill a few trees, but the Cuban pine in northeast Cuba—like the pines of the southeastern United States—is a fire-adapted species. While in the mountains we saw an area where a wild fire had swept through within the past year. Many trees had been killed—most pines had survived. The understory was nicely opened. Unfortunately, although I pointed it out to the Cubans, they were blind to the positive value of fire. The attitude I encountered was similar to the "Smokey the Bear" attitude of fifty years ago in the United States.

More Recent Searches

Since 1988, there were expeditions in search of ivory-bills in the Los Rusos area in 1989, 1991, 1992, and 1993. In April 1991, John W. McNeely may have had a glimpse of an ivory-bill in the Sierra de Moa. He visited the upper reaches of the Jaguani River and found some old-growth pines and hardwoods, including two dead pines that showed work that he believed was that of a foraging ivory-bill.[34] McNeely also searched for ivory-bills near Ojito de Agua and found none. What he did find, however, was a prospector's camp just a few hundred yards from the 1986 sightings. The search for chromium in the area was stopped immediately as a result of efforts by the Cuban Academy of Sciences to protect the habitat, but this was not accomplished before considerable damage had been done.[35]

Martjan Lammertink, a Dutch ornithologist, spent 120 days in the area in 1989, 1991, and 1993 and found no evidence of existing ivory-bills.[36] His study included efforts at Ojito de Agua; a site called Nuevo Mundo, about five miles to the northeast; and at the Jaguani Reserve sixteen miles to the east of Ojito de Agua. All of the sites had been cut over. The Nuevo Mundo site had been cut recently (mostly in the 1980s), and Lammertink believed that it had supported ivory-bills. Garrido estimated that the Jaguani Preserve had supported four to five pairs in 1987, and residents of the area remembered ivory-bills but said they could no longer be found there.[37]

Lammertink reported that none of the conservation plans recommended following the expeditions in 1986, 1987, and 1988 had been implemented.[38] A survey of potential habitats in eastern Cuba by Lammertink and Estrada[39] between February and May 1993 provided little hope. They did find almost seven square miles of untouched forest near Pico la Bayamesa in the Sierra Maestra but believed that the area was probably outside the species' range. In spite of the bleak prospects, Lammertink and Estrada (as Short and the Piciform Specialist Group of ICBP had before) recommended that the areas around Ojito de Agua and Pico de Bayamesa be fully protected.

Lammertink concluded that the birds are extinct in Cuba—which may be true, but is not a conclusion reached scientifically.[40] It is easy to knowledgeably say that something exists—all one must do is provide acceptable proof. But what kind of proof does one produce to "prove" that a species *is* extinct? The ivory-bill, in particular, is a species that has disappeared for years and has repeatedly been declared extinct, only to be rediscovered. Perhaps Lammertink's pronouncement and that of the US Fish and Wildlife Service are the stimuli needed to get someone to prove that they are wrong. With no malice toward the naysayers, I so hope it can happen.

Indeed, new observations in 1998 suggested the "probable presence" of ivory-billed woodpeckers in the Sierra Maestra in southeastern Cuba.[41] My discussions with Garrido and Arturo Kirkconnell, respected Cuban ornithologists, in July 2001 revealed that presence of the ivory-bill in the Sierra Maestra had yet to be confirmed and that their continued existence in Cuba is simply unknown. I visited the Sierra Maestra in November 2002 and saw some potential for the birds to be in remote forests of the rugged terrain. As of late 2003, organized searches in the region had produced no firm evidence of the birds.

Lammertink suggested that the Cuban ivory-bill might have been saved if conservation measures had been taken twenty-five years earlier and that it "holds the dubious honour of being the one species that was exterminated by the Cold War."[42] Certainly the Cold War resulted in keeping most scientists from seeing the birds, but we must consider that it was often US corporations that were clearing the virgin pines just before the Cuban Revolution and no doubt that clearing would have continued even quicker if the Revolution had not taken place. Indiscriminate killing of the birds for food, trophies, and simple curiosity, was—and may still be—a problem as well. This is a function of lack of education and a direct result of opening up the forests. In spite of talk of conservation and habitat protection following publicized sightings of ivory-bills in 1948, 1956, 1984, 1986, 1987, and 1988, cutting of old-growth forests was, if anything, only put on hold. It ultimately continued. For example, three square miles of mature pine forest in the area of the 1984 sighting were cut in 1992.[43] Thus, the ivory-bill is a bird that, if it is gone, is gone primarily because of efforts to maximize the economic use of the forests without regard for the diversity of life that depended on it.

CHAPTER 11

An American Icon

$\mathscr{F}$rom the late 1800s to the present, the rarity of ivory-bills has made them a focus of attention and cultural symbolism. And there is an eclectic group of modern ivory-billed woodpecker searchers with a story that is as equally fascinating as that of Native Americans, scientists, birders, and natural history collectors. These are the advertisers, conservationists, collectors, jewelers, artists, writers, and a variety of others who did and do inhabit America and the world at large. Their attraction to the ivory-billed woodpecker probably has its roots in the size, rarity, striking color pattern, and heavy, ivory-colored bill. Many considered the bill to be *real* ivory. For them the ivory-bill was a natural object of curiosity and even a prized meal for a hungry traveler. Hunters knew them, but so did the southern farmers and the enslaved generations of the American South. It was Mark Catesby's, and later John James Audubon's, portrayal of ivory-bills that introduced them to the imaginations of a wider audience, extending their range to minds outside of the southern forests. Audubon wrote that in his travels, he found that ivory-billed woodpecker heads were commonly sold at refueling stops along the Mississippi River, with two or three heads costing a quarter.[1] Some were supposedly used as watch fobs, although thus far in my searches I have not located any, nor illustrations or further mention of any.

An Icon for Advertising

As with anything large and rare (and thus capable of eliciting attention), the ivory-bill has been used in advertising. Take, for example, a 1951 Travelers Insurance Company advertisement, which featured a pair of ivory-billed woodpeckers in a full page of the May issue of *National Geographic*. Titled "The Woodchopper with the Big Appetite," the advertisement says that the ivory-bill "has a new lease on life" because a 1,300-acre sanctuary had just been set aside for it in Florida. The ad went on to note that the birds depend on beetle larvae that they dig from beneath the bark of dead trees. The tie-in with insurance was stretching things a bit, but the conservation message was not bad: "To keep eating, the ivorybill has to keep hacking away. If he runs out of dead trees, he is like the family man who has no insurance. Each is safe only so long as he can keep working." In the 1980s, one enterprising liqueur company wishing to focus on the rare achievement of its product published an advertisement in *Time* magazine showing the even more rare (at that late date) ivory-billed woodpecker that had chipped away at a large tree to reveal a bottle of its product (Plate 9). The caption was "A Grand Achievement."

Following a 1999 report of ivory-bills being seen in the Pearl River Swamp of southeast Louisiana by a Louisiana State University graduate student, the Carl Zeiss Optical Company, a maker of quality binoculars, decided to get involved in the search for the feathered Holy Grail. As discussed in an earlier chapter, the company funded searchers who spent a month in the area in the spring of 2002.[2] Whatever advertising value Zeiss got out of the effort (coverage was widespread), they may well have deserved. They did things the right way—creating an advisory committee of knowledgeable scientists (including myself) and enlisting master birders to carry out the search.

Beyond the commercial interests, the ivory-billed woodpecker has become an icon in the not-for-profit world of clubs, societies, and conservation organizations. The ivory-bill is a symbol used by the Louisiana Ornithological Society, both on the masthead of its newsletter and on an arm patch. Because the last confirmed and best-studied ivory-bills in North America were found in northeast Louisiana, some feel the state still may contain a hidden population. But for a group interested in all birds, the selection of the ivory-bill as an emblem speaks of the bird's meaning to birders. In a similar manner, the Birmingham, Alabama, Boy Scout Order of the Arrow Lodge, an elite honorary scouting organization, uses the ivory-bill as its symbol. To them, the ivory-bill is a totem, much like the totems of Native Americans. The symbolism of the ivory-bill even extends across the Atlantic. In 2001, a prominent picture of an ivory-bill was used by the British Ornithologists' Club to help raise funds for purchase of habitat in Cuba—ostensibly ivory-bill habitat if the bird is still there.

On a broader scale, the National Audubon Society used the ivory-billed woodpecker as a rallying point and as a focus for fundraising in the late 1930s and early 1940s. These efforts resulted in James Tanner's dissertation research and promoted the efforts to save the ivory-bill in what many consider its last stand in the United States, the Singer Tract. In recent years the National Audubon Society also has used the ivory-bill to obtain funds to search for the bird and promote conservation of habitat for other species in Cuba.

The plight of the ivory-billed woodpecker was one of many species that was championed by Rosalie Edge and her Emergency Conservation Committee—and that helped finance the committee's conservation efforts.[3] Edge was a strong conservation advocate who simultaneously attacked government agencies and the National Audubon Society for their failure to secure protection for the Singer Tract. Her battle with the National Audubon Society was long-standing and usually related to how the organization spent its money. In 1944, the National Wildlife Federation joined the fight to help the beleaguered ivory-bills of the Singer Tract by highlighting the species on one of its "conservation stamps." The stamp featured a painting by Walter Weber of a male ivory-bill on a stub. Other ivory-billed woodpecker National Wildlife Federation conservation stamps followed (Plate 8).

As a symbol of rarity and loss, the ivory-bill has been used for both profit and collecting for a worthy cause. But it is also a collectible in and of itself, as evidenced by ivory-billed woodpecker trading cards. Advertising trade cards became popular during the Victorian era of the 1800s, and their popularity as antiques has lasted. These small cards are often packaged with a product or simply distributed to potential customers. Each card has an advertisement on it, but also some image and text that varies from card to card. Some cards fetch hefty prices among collectors. Cards with wildlife on them have been issued for more than a century. Arm and Hammer and Cow Brand baking soda (different trade names for the same product produced by the same company) began issuing trade cards with birds on them at least as early as the 1880s. Early bird cards were often of exotic species and included no information about the birds. By the early 1900s, however, companies were reaching out to the popular conservation movement with cards that provided good images and sound biological information about North American birds, as well as an advertisement for their product. Messages often included a conservation theme, and the cards and posters and booklets associated with them were donated to schools for nature study classes.

The front of card "No. 21" of series two (1913) and card "No. 15" of series seven (1918) of the Arm and Hammer brands display the same color illustration of a male ivory-bill by Mary Emily Eaton, a British botanical illustrator (Figure 11.1). The painting illustrates the bird in a mature open-forest habitat. It is also a good example of an error often seen in woodpecker illustrations: the slant of the tree relative to the opening of the nest cavity. It appears that many illustrators like their woodpeckers placed vertically on the page, resulting in a tree and nest cavity that opens upward rather than on the underside of a limb. The upward entrance would invite the rain to fill the nest. Nonetheless, the back of the cards demonstrate the expanding concern about ivory-bills in the early twentieth century: "Once

Figure 11.1. Arm and Hammer Baking Soda trade card featuring ivory-billed woodpecker; issued in 1913 and 1918.

a common bird in the southern states, it is much to be deplored that this fine woodpecker is now threatened with extinction."

Ironically, the Singer Sewing Machine Company, the company that owned the Singer Tract where Tanner studied the woodpeckers, also issued trading cards with birds on them. However, Singer never produced cards featuring ivory-bills. The Singer cards had both birds and a picture of an egg for the species as well as a message about the bird with the advertisement, but these were abandoned as a marketing tool just as the controversy over the cutting in the Singer Tract heated up.

Jewelry, Ceramics, and Whiskey Decanters

Native Americans collected and wore the feathers and bills of ivory-bills—not as adornment but as a totem indicating or aspiring to emulate characteristics attributed to the bird. In western culture jewelry with an ivory-billed woodpecker motif has occasionally appeared. A glitzy rhinestone

Figure 11.2. A 2³/₄-inch-long, rhinestone-studded ivory-billed woodpecker brooch probably created in the early twentieth century at about the time the ivory-bill was making its last stand in northeast Louisiana. *Courtesy Jerome A. Jackson*

ivory-bill that dates from the late nineteenth or early twentieth century is one example of the woodpecker inspiring a jewelry piece (Figure 11.2). The bird not only has a jeweled body, but the bird's head and neck are also hinged to the body so that the wearer can make it "peck."

Another form of the practical arts that used the image of the ivory-bill is pottery. In the early twentieth century, the manufacturer of Tuscan China, a British purveyor of fine china, created a vase with an ivory-billed woodpecker motif. The receptacle for a plant was a hollowed tree trunk, to which an ivory-bill was clinging. Multiple series of collectible plates have featured ivory-billed woodpeckers (Figure 11.3), and there have been several different porcelain figurines of ivory-bills produced as limited edition collector's items (Figure 11.4). These have even included bourbon decanters.

Coins, Medallions, and Stamps

Ivory-billed woodpeckers have been featured on metal items ranging from keychains to collectible medallions and silver ingots, but perhaps the most interesting is a commemorative coin minted for Cuba (Figure 11.5). In

Figure 11.3. Collectible plates featuring the ivory-billed woodpecker. *Courtesy Jerome A. Jackson*

Figure 11.4. Limited edition figurines of ivory-billed wood-peckers, including two whisky decanters. *Courtesy Jerome A. Jackson*

1994, Spain minted commemorative coins as part of an Ibero–American series depicting endangered species. The series included coins for eight Latin American countries plus Spain and Portugal. Wildlife featured included the ivory-bill; the Spanish lynx (*Lynx lynx oardinus*, Spain); gray wolf (*Canis lupus*, Portugal); giant armadillo (*Priodontes maximus*, Argentina); horned guan (*Oreophasis derbianus*, Guatemala); pampas deer (*Ozotoceros bezoarticus bezoarticus*, Uruguay); Galapagos penguin (*Spheniscus mendiculus*, Ecuador); Pacific Ridley sea turtle (*Lepidochelys olivacea*, Mexico); mantled howler monkey (*Alouatta palliata*, Nicaragua); and the yellow-tailed woolly monkey (*Lagothrix flavicauda*), vicuna (*Vicugna vicugna*), and American crocodile (*Crocodylus acutus*) all on the same Peruvian coin. These silver coins were minted to call attention to the plight of endangered species as well as to tap into the lucrative commemorative coin market. The 40,000 ivory-bill coins had more than quadrupled in value by 2002.

Ivory-billed woodpeckers have been depicted on at least nine postage stamps (Plate 8). Certainly these help call attention to the plight of troubled species, but of course they also provide revenue for some nations far from the range of ivory-billed woodpeckers or even any of their close relatives. In 1985, the 200th anniversary of Audubon's birth resulted in issuance of Audubon's painting of ivory-bills on postage stamps by the Central African Republic (a 130-franc stamp), and Redonda (a 60-cent stamp), a tiny island among the leeward islands of the Lesser Antilles. Five stamps with ivory-bills on them were issued by Cuba: 1956 (a 5-peso stamp), 1961 (a 2-centavo stamp), 1962 (a 5-peso stamp with the same illustration but different coloring than the 1956 stamp), 1978 (a 30-centavo stamp), and 1991 (a 5-peso stamp). An ivory-bill stamp was issued by Tadjikistan in 1999. The United States has yet to issue a stamp featuring the ivory-billed woodpecker, but such a stamp would likely draw considerable attention to both the bird and the extensive old growth forests it needs.

Figure 11.5. Ten-peso Cuban commemorative coin featuring an ivory-billed woodpecker.

Ivory-Bill Art

Probably no image of the ivory-billed woodpecker has been reproduced more than the one from Audubon's *Double Elephant Folio*. The plate depicts a male and two females on a dead branch, the male and one female looking at what appears to be an adult "Betsy" beetle (*Popilius disjunctus*) and the other female with her tongue partially extended. This plate was number LXVI among Robert Havell's engravings of Audubon's work in the *Double Elephant Folio* and was Audubon's original plate 181. Dozens of editions of Havell's engravings of Audubon's work have been issued over the years, and prints of this ivory-bill are readily available. The original painting is in the collections of the New York Historical Society.[4]

The best collectible art prints of the ivory-bill, in my judgment, are those lifelike paintings of Guy Coheleach (Plate 7) and Julie Zickefoose and the stylized geometric art of Charles Harper. Each of these demonstrates an exceptional understanding of the behavior and ecology of the species in addition to the superb technical skills of the artist. In the first two cases, each artist has effectively portrayed some aspect of the biology of ivory-bills that had previously been only portrayed with written descriptions.

Ivory-Bills in Literature

Literature also abounds with references that are unquestionably or likely references to ivory-bills. It seems highly likely that rural African Americans in the American South knew of ivory-billed woodpeckers, and it also seems likely that there were unique cultural ties linking African American culture to the ivory-bill. One of these might be found in William Faulkner's story "The Bear."[5] In the story, Faulkner refers to a big woodpecker called "Lord-to-God" by the "Negroes." In the fall of 1988 I was quite excited to see a display of a new book at a local bookstore: *Deep Enough for Ivorybills,* by James Kilgo, an English professor at the University of Georgia.[6] No question about it, I had to have a copy. Kilgo, it turns out, knows the swamps along the Savannah River and the low country of South Carolina—ivory-bill habitats, of course. Much to my disappointment, it turned out that the book is not about ivory-bills but about hunting and fishing and human relationships. But Kilgo's descriptions get one into the swamps, and his drawing of ivory-bills at the front of the first chapter is excellent. He is obviously a keen observer of nature, even if our bird did not win the starring role the book's title implied. Another novel in which the ivory-bill gets a cameo role as a symbol of hope is Walker Percy's apocalyptical novel *Love in the Ruins.*[7] Set in Slidell, Louisiana, Percy describes the Honey Island Swamp and environs—the same area where the Carl Zeiss Optical Company sponsored the ivory-bill search.

A curious thing happened in 1993. Within a matter of about six weeks I received telephone calls or letters from three different fiction writers seeking information about ivory-bills and wanting to discuss the birds as potentially being central to a mystery novel. I think two were particularly fascinated with the possibility of intrigue relative to the bird's presence in Cuba. They all saw the potential for a good story. Certainly this bird is a mystery. Thus far I have not seen an ivory-bill novel by any of these authors, but in 1996, yet another author, Greg Lewbart, tapped into that literary potential.[8] Lewbart, a veterinarian, worked at a clinic in Naples, Florida, not far from the Fakahatchee Strand and some of south Florida's wildest country. There have been rumors of ivory-bills in the area for decades, and Lewbart saw the area as the perfect setting for a story focusing on the accidental discovery of nesting ivory-bills by an unscrupulous entrepreneur. The GPS (global positioning system) coordinates for the location of the nest are sold to one person who wants the birds for a private collection and also to a biologist who sees fame and fortune in studying them. The story has all the elements of a good read: greed, murder, intrigue, adventure, and a certain level of plausibility. The latter quality came as a result of Lewbart's veterinary background, familiarity with the setting, and sound research regarding the birds.

Perhaps my favorite literature reference to the ivory-bill is a children's poem written by Elizabeth Gordon and published in 1912 in a book titled *Bird Children* (Figure 11.6).[9] The poem is simply titled "Ivory-Billed woodpecker," and it strikes me as more than a children's poem:

> *Ivory billed Woodpecker said:*
> > *"Dear me!*
> *They're cutting down my family tree;*
> > *Where can I live, I'd like to know,*
> *If men will spoil the forest so?"*

There is a rich history of human association with ivory-bills, and there is a certain folklore as well. For example, there is the true story about Mason Spencer told in Chapter 6.[10] Spencer was told that ivory-bills do not exist. Then he shot an ivory-bill and presented it to the Louisiana Department of Conservation to prove them wrong. This story, however, has taken many different forms. Spencer really did visit the Louisiana Department of Conservation and was told that ivory-bills were extinct. He really did believe they existed. He really was given a permit to shoot one as proof. And in 1932, he did. The specimen exists. But other versions of this story have appeared in the literature as if they were true, with no substantiating evidence. The major differences are the times and places where the events took place. Stories of murdered ivory-bills abound, and I will relate only a few of these.

IVORY BILLED WOODPECKER said:
"Dear me!
They're cutting down my family tree;
Where can I live, I'd like to know,
If men will spoil the forest so?"

Figure 11.6. Ivory-billed woodpecker drawn by M. T. Ross from Elizabeth Gordon's book *Bird Children,* 1912. Clearly Gordon's message shows that by 1912 we understood what problems the ivory-bills faced.

Wildlife artist Donald Eckelberry wrote of one:

[The] Louisiana State Conservation Department . . . forbade the collecting of any more birds; however, in order to make their rarity and value appreciated by the American public, a newspaper story had put an estimated value of one thousand dollars on an ivory-bill skin. Forthwith a young fellow appeared with one of the big woodpeckers dangling from his hand, to collect the one thousand dollars. Wardens in the Singer Tract thereafter kept stricter vigil.[11]

In 1956, another ivory-bill was reported shot in North Carolina. That shooting was reported first in the *Philadelphia Inquirer* (April 11, 1956) and then reprinted in *Audubon Magazine*.[12] No evidence was ever presented that the shooting actually happened, and the National Audubon Society never followed up on the report.

A story with substantial dialogue comes to us from Texas. Archer Fullingim, publisher of the *Kountze News* in Kountze, Texas, the heart of the Big Thicket country, tells the following story of an event that happened about 1961:

About six years ago, I published a picture of the Ivory-billed Woodpecker in *The Kountze News,* saying it was extinct over all the nation except possibly in the Big Thicket, and that there might be a few left in the deep recesses of the tight-eye Thicket.

About a month later, a native of the Thicket came into the News printing shop carrying a dead woodpecker. "Ain't this the bird you said was extinct?" he asked.

At once I thought it was, but did not say so, and hunted up all the pictures and descriptions I had of the Ivory-billed Woodpecker, and I read the detailed description to the hunter. As I read he checked off each identifying point on the dead bird.

"Hit's hit all right," he said, "I studied this particular woodpecker for a week before I shot one. I got in real clost. I can't read, but I had my boy read that what you put in the paper 'bout the woodpecker, every night. I memorized hit. I killed hit because I knew you wouldn't believe me unless I tuck hit to you."

My face must have revealed my utter despair, frustration, and sadness. He looked at me intently and finally grinned.

"You needn't feel so bad about me killing this un. There's a whole lot more where I got him. But I ain't goin' to tell nobody. Not even the feller from A&M; and I ain't going to tell you where hit is. In fact, I'm goin' to take this bird back with me. If I've got all them extinct woodpeckers, I'll be the most important man in the world, even if nobody knows hit but me, and that's the way I want hit." [13]

In the book, *The Big Thicket. A Challenge for Conservation,* author A. Y. Gunter told the same story about Fullingim's experience with a slightly different twist, and as occurring about seven years later, in 1968:

Around four years ago a grizzled backwoodsman walked into the office of Archer Fullingim, editor of the *Kountze News.* The man asked Archer whether he thought there were any of those Ivory-billed birds like the ones whose pictures had been run in the newspaper. When Archer replied skeptically the man reached around and dumped a dead Ivory-billed Woodpecker on the editorial desk. The editor of the *Kountze News* stared in dismay and unbelief. "Those birds are out on my place," the settler grinned. "They're the rarest birds in the world, and I've got 'em. Ain't nobody but me know where they are." The man pocketed the bird in his jacket and walked out, remarking that there were "plenty more where that one came from." Fullingim learned later, however, that the man had seen no more Ivory-bills. They deserted the man's land, seeking, no doubt, a deeper solitude somewhere in the Thicket.[14]

Another bit of folklore relating to the ivory-bill came from my interview with ninety-year-old Stella Christopher of Warrensburg, Missouri. She told me of watching ivory-billed woodpeckers in a cypress swamp near her home back in Red River Parish, Louisiana. She noted that the "Negroes of the area said that the tapping of ivory-billed woodpeckers was saying to them 'Good God, fish here'"—a mnemonic clearly fitting the characteristic double rap of ivory-bills.

The stories continue, as does the fascination with them. Perhaps they are simply tales too good not to be told, but sometimes a grain of truth lies within a good story. Certainly there is bound to be confusion: mistaken identity involving ivory-bills and other woodpeckers. But if they are out there, they have probably been seen by someone, and well may have been seen by some who did not realize the significance of the sighting. There is no doubt that even a confirmed sighting will, over time, be transformed in folk tales. And in this way the ivory-bill further enriches the culture.

Cultural Encounters

There are, of course, tales of weird cultural encounters with ivory-bills. I once visited the small collection of birds in McBride Hall at the University of Iowa because I had heard they had two ivory-billed woodpecker specimens—and they did. But in the course of digging through accession records, I found evidence that there had been a third ivory-bill specimen. When I asked curator George Schrimper about the third one, he recounted a bizarre story. One night in late February 1979, a student had thrown a brick thorough a glass display case and had taken the specimen—a female ivory-bill collected in Volusia, Florida, in February of 1869. The thief had later mailed a photo of himself (wearing a mask) with the bird to the state newspaper, the *Des Moines Register*.[15] He admitted the theft, but wrote that he would not return the bird. His reason? He loved it! He said that his theft of the specimen was a result of "genuine love for the bird" and that he was not "an agent for some radical bird group . . . not a violent man . . . not a political entrepreneur." He planned to take the ivory-bill on a tour of the Gulf states where, together, they would seek information on reports of living ivory-bills. The student's anonymous letter to the newspaper ended with the promise that he would "honor" the ivory-bill "with all the love and care it deserves."

Figure 11.7. Tattoo on the leg of Kate Somerville: the last living ivory-bill? *Courtesy Kate Somerville*

Authorities tried to obtain the letter and photo, but, according to Schrimper, were denied access to them, citing the First Amendment of the US Constitution and confidentiality of newspaper sources. The ivory-bill specimen was never recovered.

Perhaps the only living icon, and, if you stretch things, the last "known living ivory-bill" was one I found a few years ago. I was asked to present a program on ivory-billed woodpeckers for the annual meeting of the Delaware Valley Ornithological Club. The meeting was held at the Philadelphia Academy of Natural Sciences and was well attended. After the meeting a young lady named Kate Somerville came up to me and we spoke extensively about ivory-bills. She was absolutely enthralled with them. She enthusiastically asked me to send her some pictures of the birds. I did. A few months later I received an envelope from Kate. It contained several photographs of a beautiful tattoo of an adult male ivory-bill drawn on her leg (Figure 11.7). In her letter to me she sadly, but proudly, commented: "Now I own the only living ivory-bill." And so, the ivory-bill woodpecker, perhaps destroyed by humanity, also lives on through our limitless imaginations and imperative to collect.

Epilogue
The Truth Is Out There

$\mathcal{T}$he epilogue title is fitting. This is a quote that often appeared on television screens at the opening of the now "extinct" but popular television series, *The X Files*. As aficionados know, the series focused on the exploits of FBI agents seeking proof for the existence of extraterrestrials on earth. At least one agent, Fox Mulder, was convinced of the existence of extraterrestrials, but he never quite seemed to have the definitive proof. He was often ridiculed and criticized for his efforts. Such is also the lot of a biologist who "believes" in ivory-bills. Individuals who think they have seen or heard an ivory-bill are often quickly dismissed and told that they are mistaken (or even characterized as crackpots). As a result, ornithologists and others who entertain the notion that ivory-bills might be "out there" are often reluctant to share their observations.

To be sure, most reports are in error. Many are ignored or go unnoticed. A few, such as the possible sighting by David Kulivan in the Pearl

River Swamp of Louisiana in 1999 attract media attention and become well-known. Some are published in obscure places such as a chapter in Fielding Lewis's book *Tales of a Louisiana Duck Hunter*.[1] Such reports often only come to the attention of biologists long after the fact, yet remain in place as a result of their publication to be debated and to become "evidence" of the ivory-bill's existence. The similar pileated woodpecker is a constant source of confusion, and other birds—and even frogs—can make vocalizations that are mistaken for the calls of ivory-bills. A major problem is that the word is out about what to say and what not to say if you want an ivory-billed woodpecker report to sound credible. I have received numerous "sightings" that quote nearly verbatim the ivory-bill descriptions from field guides, news articles, and other published sources. Or do they? Could a sighting be so close to what the textbooks say it could be that the person recounting the experience gets it just right? An individual can take great satisfaction in a "good" sighting, but the scientific community can never know for sure from a verbal report—which is why we must have solid evidence in hand from any observer or observers before science can claim that ivory-billed woodpeckers still exist. The probability of ivory-bills still existing in the United States or Cuba is slight—but there is a possibility. While birders seek the Holy Grail, however, others see the ivory-bill as a problem.

In 1985, the US Fish and Wildlife Service initiated a status review of the ivory-billed woodpecker to determine if it was extinct and should be removed from the Endangered Species list.[2] Results of that review have not been released. On the political front, an organization calling itself the National Wilderness Institute in 1997 petitioned the US Fish and Wildlife Service to remove the ivory-bill from the Endangered Species list (those species protected by the Endangered Species Act) because it has long been extinct. They argued that having the ivory-bill on the Endangered Species list was costing taxpayers $37,000 per year—although they did not say how. As of May 2003, that petition has not been acted on.

With continuing search efforts, and with the combined abilities of dozens of competent ornithologists and birders, why can we not say for certain that the ivory-billed woodpecker is gone forever? It is complex. The biology of the ivory-billed woodpecker, the attributes of its habitat, and human nature prevent a declaration of extinction from being scientifically defensible at this point. There just might be ivory-bills out there.

Inherent in the biology of the ivory-bill is its wariness of humans—perhaps a trait selected for by hunting pressure from those who would eat it, or collect it for its bill or plumage, for profit or for science. It has never been an easily approachable bird. Arthur T. Wayne, one of the most successful collectors of ivory-billed woodpeckers during the waning years of the nineteenth and early years of the twentieth century, made extensive forays into South Carolina swamps to collect such rare swamp dwellers as Bachman's warbler. When writing about the ivory-bill, he noted, "I have thoroughly explored, during the past twenty-five years, nearly all the great swamps from Charleston to the Savannah River with the hope of finding it."[3] Yet despite the fact that ivory-bills were apparently present less than forty miles north of Charleston, he failed to locate them. And recent efforts in Cuba by Cuban scientists and by Lester Short, Ted Parker, George Reynard, Jennifer Horne, John McNeely, and myself have produced reports of brief glimpses and sounds of ivory-bills—but none of the documentation required by science to *prove* that they are there.

This wariness, combined with the ivory-billed woodpecker's sometimes quiet nature, make it a difficult bird to find. Take, for instance, the following results of one field trip made by Arthur Allen and Peter Paul Kellogg:

> failure to find the birds in a given area is no proof that they are not there, for they are not noisy except when disturbed; their voice does not carry nearly as far as that of the Pileated Woodpecker and in the big trees they normally frequent they are easily overlooked. We camped for five days within three hundred feet of one nest and, except when the birds were about to change places on the nest or were disturbed, seldom heard them.[4]

The two had searched for these birds for three days without hearing them, although they had frequently been within three hundred yards of the nest. They finally located the birds because they just happened to be within hearing distance when the pair vocalized as they exchanged places at the nest.

Several ornithologists have emphasized the lack of carrying power of the voice of the ivory-bill relative to that of the pileated woodpecker. In 1898, Robert Ridgway noted the ivory-bill "is comparatively quiet and secluded, and its notes would not attract attention except from one keenly alert for new sounds."[5] The birds could be noisy, but the nature of ivory-bill vocalizing varies greatly depending on many factors, and at times they could be inconspicuous.

I believe the ivory-bill has an elaborate social system or at least a long juvenile dependency, possibly even a cooperative breeding system similar to red-cockaded woodpeckers. John Dennis reported three adult birds at a nest in Cuba, and many authors have reported the birds traveling in pairs or family groups.[6] No other *Campephilus* woodpecker is known to have such a breeding system, but then no other *Picoides* woodpecker has a system like that of the red-cockaded. If ivory-bills disperse as a family unit, this would allow opportunistic breeding and thus enhance the potential for the species to have survived to the present.

Another important aspect of the ivory-billed woodpecker that might allow for its continued existence is the probability that the birds are long-lived. In general, larger birds live longer than smaller ones. So given that some of the more common small woodpeckers, such as the red-cockaded, are known to live as long as seventeen years, it is conceivable that ivory-billed woodpeckers could have a natural potential longevity of 20-plus years. Such longevity would increase the probability of birds surviving to the present. Birds sighted in the Singer Tract in the 1940s might well have lived into the 1960s.

Small numbers of birds alone would be factor enough to make their discovery difficult. Add to that the large size of their home range and the inhospitable nature of their habitat and what James Tanner referred to as the problem of finding "an animated needle in a haystack."[7]

The kinds of bottomland habitats in which ivory-billed woodpeckers have been found are flooded for several weeks out of the year, and even during the dry season the hiker is faced with hiking on dry land one minute and then having to cross a shallow body of water the next. The nuisance factors of poison ivy, mosquitoes, ticks, poisonous snakes, an occasional wild hog (*Sus scrofa*), and hunters are discouragement enough for all but the most dedicated. Most people who visit southern swamp habitats do so only during the hunting season or for purposes of clearing or filling them. Even hunters restrict their visits to the most accessible areas. Birders typically watch from the road, a bridge, or a canoe on open water. They rarely penetrate more than a hundred yards into a forest unless there is a road or trail, making chances of an ivory-bill sighting slim.

One of the most encouraging aspects of my searches for ivory-billed woodpeckers is that many prime areas are already under some sort of protection, a result in large part to efforts initiated since the species was studied in the Singer Tract. This protection has paid dividends in restoring mature forests destroyed by indiscriminant cutting of the past. Increased interest in conservation in recent years has resulted in establishment of the Big Cypress National Preserve (1974, Florida), Big Thicket National Preserve (1974, Texas), the Pascagoula Hardwoods Preserve (1976, Mississippi), and the Tensas River National Wildlife Refuge (1980, Louisiana).[8] Other areas, such as the Okefenokee National Wildlife Refuge (1937, Georgia) and portions of the Santee Swamp in Francis Marion National Forest (1936, South Carolina), have been protected for more than half a century. Added to these are portions of several other national forests (e.g., Ocala, Apalachicola, Osceola; Florida), other national wildlife refuges (e.g., St. Marks, Florida; Noxubee, Mississippi; Savannah River, Georgia), portions of several military bases (e.g., Fort Stewart, Georgia; Eglin Air Force Base, Florida), numerous state wildlife management areas, and sanctuaries owned by conservation organizations such as the National Audubon Society (e.g., Silver Bluff, South Carolina; Corkscrew Swamp, Florida). The bottom line is there are several hundred thousand acres of swamp forest that are in various states, from essentially virgin to young regeneration. Because of their aquatic nature, few are truly isolated and many are more or less linked like pearls on a string by the river systems that nurture them.

This linear habitat dispersion facilitates successful dispersal of birds into any remaining suitable habitat. Although channelization has wrought enormous negative changes in our riverine ecosystems, a fringe of bottomland forest still exists along most streams. Such ribbons of habitat could serve as corridors for birds dispersing from one larger forest tract to another. The distributional pattern for ivory-billed woodpeckers has always had a linear component as a result of the species' habitat preferences.

So what happened to the birds in the Singer Tract? It is possible that they lived their lives out and then died there, but it is equally possible that they dispersed following disruption of their habitat. The most logical direction of dispersal would have been downstream toward the Atchafalaya Basin, but it is also plausible that they went east to the Mississippi and then followed any of its major tributaries. The Yazoo River, for example, is only about thirty miles from the Singer Tract.

Another major category of reasons the ivory-bill may have eluded us for so long can be found in human nature. Most individuals would not recognize an ivory-billed woodpecker if they saw or heard one, and even if they did, the odds of them reporting it are slim. I was particularly impressed with this problem during my conversations with biologists responsible for managing some of the more promising areas. I spoke with only one who knew what an ivory-billed woodpecker sounded like, and most were not certain how the ivory-bill and pileated differed in appearance. Most thought that bill color was the key identifying feature or that ivory-bills were so much larger than pileateds that they would immediately see the difference. Perhaps as important as everything else combined is the negative attitude that is prevalent: there are no more ivory-bills. If they became extinct long ago, why bother to look?

As Dennis aptly wrote,

> No one wants to experience the kind of inquisition that awaits anyone who does not have incontrovertible proof of a sighting. In former days proof lay in displaying a freshly killed specimen. Although that kind of proof is now unthinkable, standards for accepting the presence of a nearly extinct species are as rigid as ever, a fact that has led to an unfortunate impasse. If no qualified person will accept your testimony, why go to the trouble of searching for the rare species? Second, if the rarity were found, why risk one's reputation by reporting it? Although novices, who continually confuse the Pileated Woodpecker with the Ivorybill, still report sightings, today few serious bird students would dare state that they had seen or heard an Ivorybill.[9]

And if a serious, knowledgeable birder were to make a report of an ivory-billed woodpecker sighting to a scientist or government authority, it would more than likely be met with a reluctance to give the report serious consideration or a prompt responsive search.

Since the last verified reports of ivory-billed woodpeckers in the Singer Tract, there have been hundreds of sightings that have been reported from throughout (and outside of) the species' known range. Dismissing those that come in from unknown individuals of unknown ornithological backgrounds, there are still dozens of sightings by what can be considered competent bird observers. Herbert Stoddard, who knew ivory-bills in his youth, noted in his 1969 memoirs that he had "observed three ivory-bills in the Southeast in the last fifteen years." More important, he noted, "I feel quite confident that the ivory-bill still has a chance to survive."[10]

Perhaps we can dismiss the photographs that George Lowery presented to the ornithological community. Perhaps we can dismiss the sightings reported by Whitney Eastman. Perhaps we can dismiss the sightings by Dennis. Perhaps we can explain away Dennis's tape recordings that were analyzed by John William Hardy at the University of

Florida. Maybe the recording made by Reynard is not of an ivory-bill. Perhaps we can dismiss the response to tape recordings that were heard by Robert Manns, Malcolm Hodges, and myself, or the birds heard by Fred Sibley and Ted Davis. But the list goes on and on—right up to the present. If each of these observations has any probability at all of having been real, these probabilities add up. Is it likely that *all* of these reports are misidentifications?

Another problem in the search for the ivory-billed woodpecker is lack of funding for researchers. Being that the world of the researcher is so often ruled by grant-based projects, there is a tendency for researchers to put their efforts into projects with a visible, definite payoff—especially because administrators seem to have a predisposition to award grants on the basis of potential for positive results and to reward researchers on the basis of published research. Searching for ivory-billed woodpeckers is not seen as having a high probability of success and it is not likely that journal editors will be chasing down the scientist to publish the results of a failed ivory-billed woodpecker search. Without the promise of funding, a truly thorough search is not likely.

In spite of repeated, and to some extent continuing, conservation efforts, and in spite of hundreds of thousands of acres of forest land now protected, we have no known population of the ivory-billed woodpecker. Why?

The answer is, of course, complex. But it can be simply summarized as a lack of understanding and a lack of commitment. These deficiencies that have led to our conservation failures have often been byproducts of the slow wheels of bureaucracy, the strident pace of capitalism, and the exigencies of war.

Yes, the Singer Tract is now protected—but we closed the barn door forty years after the horse got out! It goes deeper than that, however. Setting aside a forest for wildlife does not necessarily provide for the wildlife that would have occurred there under natural conditions. Any given forest stand is affected by the habitat that surrounds it. Closing the doors on a piece of property to "preserve" it will not necessarily conserve it. Although the controlling physical characteristics of upland and bottomland forests are quite different (e.g., fire in the uplands and flooding in the lowlands), the ecological impacts of thwarting the natural functioning of these environmental factors often have similar impacts on the plant and animal communities the forests support.

The uplands of the southeastern United States were naturally burned with increasing frequency as one nears the Gulf Coast or moves down the Florida peninsula. The fires were, and still are, started by lightning and carried by pine straw and other litter on the forest floor. They are normally "cool" fires, moving quickly through the forest as a result of the volatile resins in the pine needles.

A cool fire may sound like an oxymoron, but imagine sitting at a candle-lit table in a restaurant waiting to be served. The dinner is taking longer than usual and you casually pass your finger back and forth through the flame—a kid's parlor trick. Now try holding your finger in the flame. Of course you would not do this, because it would burn. Fire moving through a pine forest is akin to your finger passing through the candle flame—the movement prevents serious damage. Such fires open up the forest, allowing sunlight to reach the forest floor, promoting growth of young pines and other fire-adapted vegetation.

Natural fires at times burned through thousands and thousands of acres. With the arrival of European humans, however, things changed. We built roads—every one of which acts as a fire break and limits the area of a fire's impact. We actively extinguish fires in the name of protecting human lives and property. We have also favored commercially important species over those of lesser commercial value. The result affects microclimates, plant and animal species diversity, the dispersal of creatures and consequent gene flow among populations, and the nature of wind-blown pollen from areas surrounding a designated wilderness. The key is that unless a wilderness area is incredibly large— at least hundreds of thousands of acres—locking the doors to it will not ensure that the ecosystems included within its borders will function normally.

Equally to the point, those areas that we do protect today are often managed to "promote forest health"—we harvest trees at young ages so we do not have dying or dead trees around. When we have an insect epidemic, we salvage infested trees. Such actions do *not* promote the health of forest ecosystems. In many ways a dead or dying tree is more important to the health of a forest ecosystem than a living tree. Decaying wood becomes soft enough as a result of fungal activity that large beetle larvae can consume it, woodpeckers can make cavities in it, and literally dozens of other species find refuge and nest sites in the woodpeckers' cavities. The process of decay is like a time-release vitamin pill, slowly releasing nutrients back to the forest. Instead, under commercial forestry practices, we remove them. (Commercial forestry practices are in general use in most national forest lands and many national wildlife refuges. National and state parks often thwart natural ecosystem processes by excluding fire and by constructing roads.) The natural functioning of these forests is important to the ivory-bill as well as to other species, endangered and not.

Bottomland forests—the heart of ivory-bill country—suffer because water is not allowed to play its natural role in the bottomlands. Periodic flooding was a natural process that shaped our bottomland forest ecosystems. As Tanner described, we find different plant communities in the first bottoms than in the second bottoms, and so on. Dams, channelization, canals, and other efforts to control water have changed the timing and extent of flooding. The natural forest communities of southeastern bottomlands are grossly different than they were when the Pilgrims first arrived. To this gross alteration of the functioning of the natural bottomland ecosystems, we can unfortunately add the same litany of problems faced by the uplands: fragmentation, isolation, removal of dead and dying trees, emphasis on a few commercial tree species, and so forth.

When there have been conservation efforts, they have always been in the realm of crisis management rather than crisis prevention. We recognize problems, but for one reason or another, do not act on them until it is too late. Scientific and conservation groups have been as guilty as others. The Committee on Bird Protection of the American Ornithologists' Union, for example, noted, "[R]eports persist of recent occurrences of ivory-billed woodpeckers in the Big Thicket area of east Texas, near Ocala, Florida, and in western Louisiana. Attempts to verify a sight record of an ivory-billed in South Carolina failed, but the circumstances seemed to warrant an attempt to secure protection for the area."[11]

This was in 1964—three years before Dennis's "discovery" of ivory-bills in the Big Thicket, seven years before Norton Agey and George Heinzmann's report from central Florida, and seven years before the photographs were taken that Lowery presented to the ornithological community.

I know. I know. You cannot tie up thousands of acres of land on the basis of a report that most people do not believe. You need proof. Unfortunately, that "proof"— if it ever comes—usually comes too late. The truth is that there was little official response to the reports of ivory-bills—no concerted efforts at the time of the reports to verify them. Such efforts, I have been told by US Fish and Wildlife Service administrators, are "wild goose chases" that they cannot justify spending money on. They have to have something to show for their money. If it is not a sure bet, forget it.

I argue that intensive searches for ivory-billed woodpeckers could—and should—be an integral part of intensive studies of the biodiversity and functioning of some of the last remaining unbroken expanses of mature forest ecosystems in the United States. Such studies would have multiple end products that would help us understand the complexity of interactions and interdependencies and how these influence the ecological, aesthetic, and economic values of these ecosystems.

Where should continued search efforts for ivory-bills be conducted? Bearing in mind the problems in finding ivory-billed woodpeckers, continued efforts in the southeastern United States should focus on the largest, most remote areas of mature bottomland forest habitats within the historic range of the species. In my opinion, based on existing habitat and its remoteness as well as on undocumented reports that cannot readily be discounted, the most likely states in which ivory-bills might still exist are, in order of likelihood: Florida, Louisiana, and Mississippi. South Carolina, Georgia, and Texas would be next, and Alabama and Arkansas would be behind them. Other states, such as Missouri and Tennessee, would be highly unlikely homes for ivory-bills in today's world. But one never knows.

To ensure the highest chances of success, there must be positive education about the characteristics of the bird and ways to distinguish it from pileated woodpeckers. Most previously searched areas are state or federal lands, all of which are frequently visited by federal and state wildlife biologists. The first order of business should be to make certain that the biologists frequenting these areas recognize a potential for ivory-billed woodpeckers still existing and know distinguishing plumage characteristics and vocalizations. Because of the wariness of these birds, the best hope for finding them may lie with a chance hearing of their vocalizations and the distinctive double rap. Merely hearing the birds, however, is not conclusive evidence of the presence of ivory-bills. A good recording of vocalizations provides better evidence, although I contend may still not be conclusive. Recordings of the double rap are more problematic because of their simplicity, the possibility of their production by a pileated woodpecker or even a mechanical source, and the lack of comparative material.

Other signs of evidence can also be important, but not conclusive, and pileated woodpeckers or other creatures—including humans—can produce most. One of the supposed telltale signs of ivory-billed woodpecker feeding activity is extensive scaling of bark from recently dead trees, yet other woodpeckers, especially the pileated, can do this. Extensive scaling can also be caused by hunters who use a type of deer stand that clamps around the tree and is pushed higher and higher as the hunter climbs. Each time the hunter puts his or her weight on the stand, it bites into the bark, holding the hunter at that level. Each of those bites loosens bark, often causing it to fall away from the tree. It also often leaves a distinct horizontal line where it had taken purchase on the tree—something that needs to be looked for when trying to interpret the cause of scaling. The ivory-bill's blows are at an angle downward to the right and left, leaving more of a chevron pattern.

The ivory-billed woodpecker may be extinct. But it has been pronounced extinct before only to be rediscovered years later. Armchair declarations of extinction based on lobbying efforts do not make it so. Nor do the accumulated negative reports of weekend sorties to the swamp edge. A thousand reports of ivory-bills without acceptable documenting photographs or undisputed sound recordings are equally inconclusive. If Lady Luck is really smiling, an ivory-bill might be seen by a weekend birder, but the best chances for success will be with well-thought out, thorough, intensive searches.

My efforts, sponsored in part by the US Fish and Wildlife Service, covered extensive areas. But logistics, lack of staff power, and other problems still limited coverage. Intensive searches involving multiple observers, such as the 2002 month-long search sponsored by the Carl Zeiss Optical Company in the Pearl River Swamp in south Louisiana, are needed. We need more and lengthier efforts by well-trained, dedicated observers.

Trying to find ivory-billed woodpeckers is fraught with problems. But if ivory-bills are found, we would have other problems: ethical ones. To announce the documented discovery of ivory-billed woodpeckers in a specific location would almost ensure their demise before conservation efforts could be put in place. There is probably not much of a threat from collectors with guns, but "collectors" with binoculars who want only to add one more species to a life list could be equally lethal. The Zeiss search, for example, was preceded by considerable publicity and followed closely by the media.[12] The confirmed discovery of ivory-bills would immediately be known to the world.

Considering the large home range of ivory-bills, it is likely that any birds found would be using lands under multiple ownerships. What would be done? What could be done to secure protection for the birds? A thousand-acre preserve such as set aside in the 1950s is inadequate for even a single pair.

I wish we had such a problem. For now we can dream. We can contemplate the possibilities of hanging onto the magnificent ivory-bill and use those possibilities as reasons for protecting, restoring, and linking together natural old-growth forests of the Southeast. If we cannot save the ivory-bill, perhaps we can save other species who shared its ecosystem.

A strong conservation movement and good faith alone are not sufficient to save an endangered species from extinction. Strong laws and enforcement of those laws are also important. But laws only buy time and are subject to change as society and political fortunes change. Over the long term, the most important protection for endangered species and their ecosystems is societal change through education that teaches appreciation and changes attitudes to favor conservation. Without understanding and appreciation of the species and its interactions within its ecosystem, conservation measures may only forestall extinction at human hands. The downward plunge may only be lessened to a gentle, but none-the-less downward, spiral.

Some have argued that times have changed, that we should forgive the excessive collecting and excessive habitat destruction of the past. Things were different in the 1800s and the early 1900s, but even then some could see that excesses were excesses. S. C. Clarke of Marietta, Georgia, writing in the popular magazine *Forest and Stream* in 1885, demonstrated that there were those who recognized part of the problem. Speaking of the ivory-bill, he noted, "I have no doubt this bird is to be found in the cypress swamps of southern Florida, but nowhere within sound of the guns of the game butchers who infest Florida in winter. It is always rare, inhabiting the most solitary places remote from mankind."[13]

Can we save the ivory-bill? By as early as 1945, Tanner did not think so. But if asked if humans might walk on the moon in twenty years, a scientist might then have said, "It's not possible." Humankind can do great things. I want to believe. If there is habitat, there is hope. If there are ivory-bills out there, there is hope.

The truth is out there.

Appendix 1
A Geography of Extinction:
Where the Ivory-Bills Were

$\mathcal{T}$his state-by-state summary of distributional records for the ivory-billed woodpecker is intended to present the documented distribution of the species, along with commentary relative to the strengths and weaknesses of earlier presented "facts." Recent unsubstantiated sight and sound records are discussed in Chapter 9. Cuban records are discussed in Chapter 10.

What is a valid record of an ivory-billed woodpecker? A specimen in hand used to be the norm for record validity, but this is not an acceptable way to validate the existence of any endangered species. In addition, locality records and other data associated with some specimens are suspect. Local hunters often shot birds and sold them to collectors who may never have known the collection locality. Some ivory-bill specimen localities may be the location where the bird was obtained from the hunter, or perhaps even where it was shipped from to a northern museum. A good quality photograph could be a valid record, but a series of photos, some showing the bird in motion, would be better, and video of a living bird would be the most acceptable of records. A good quality sound recording of an ivory-bill vocalization could also be a valid record, but this is fraught with problems of possible blue jay mimicry. A sound recording of the double rap of an ivory-bill might provide evidence, but possibly not convincing validation for an ivory-bill record. Records before James Tanner's work in the Singer Tract were usually validated by specimens, but scientists also generally accept the earlier sight records of ornithologists who knew the birds as a result of previous experience with them.

Alabama

Ivory-billed woodpecker records from Alabama are all from riverine forests in the western and southern counties.[1] Five specimens are mentioned in the literature as being from Alabama, two from Dallas County and one each from Marengo, Hale, and Pike Counties, but none is known to have survived, although it is possible that the dateless specimen of a male on display at the Anniston (Alabama) Museum of Natural History is an Alabama bird.[2] Apparently acceptable early records from the literature include sight records from Lamar and Wilcox Counties.[3]

Arkansas

Records of ivory-billed woodpeckers for Arkansas date back to John James Audubon's journey down the Mississippi in 1820.[4] Douglas James and Joseph Neal summarized nineteenth-century records from along the Mississippi, Arkansas, and St. Francis Rivers.[5] Although there are at least five specimens labeled as having come from Arkansas, none has locality data sufficient to indicate county of origin. Two have labels indicating "St. Francis River, Arkansas"; others have no locality data. Apparently acceptable literature records of sightings come from Jackson, Poinsett, Mississippi, and Phillips Counties.[6]

Florida

By far the greatest numbers of both written records and ivory-billed woodpecker specimens are from Florida. This is undoubtedly in part because of the accessibility of Florida to eastern establishment ornithologists of the late 1800s and early 1900s—because they are the ones who established the records. They are also the ones who established the market for ivory-billed woodpecker specimens.

Florida was the North American stronghold of the ivory-bill because of its warmer climate and original expansive mosaic of swamps and uplands. Ivory-billed woodpeckers are known from the panhandle to the Everglades, and from Merritt Island on the east coast to Cedar Key on the west.[7] Areas figuring prominently in knowledge of the Florida distribution of ivory-billed woodpeckers include (1) The Chipola/Apalachicola River swamps; (2) the lower Suwannee River and adjacent California Swamp of the "Big Bend" area (several specimens); (3) the Big Cypress Swamp of south-central and southwest Florida; (4) numerous areas in central Florida, including swamp forests associated with the Oklawaha, Wekiva, Withlacoochee, Aucilla, St. John's, Indian, and Crystal Rivers.[8]

In 1879, W. E. D. Scott, collecting along the Withlacoochee River near the Gulf Coast, noted "the ivory-bill was by no means rare."[9] By 1881, C. J. Maynard believed the ivory-bill was "common in only one section of Florida, 'the Gulf Hummock,' [= Gulf Hammock, near the mouth of the Waccasassa River east of Cedar Key] where the birds were 'quite numerous.'"[10]

Philip Laurent spent March of 1887 collecting birds in Levy County, in the Gulf Hammock area.[11] He euphemistically noted that during that one month, five ivory-bills "came into my possession." Unfortunately he had shot them with a 44 Winchester that so blew the birds apart that only two, a male and a female, could be salvaged as specimens. He commented that the ivory-bill was "not as rare as many think" and that "all the residents with few exceptions are acquainted with it."

T. Gilbert Pearson noted that in 1886 or 1887 he had seen the heads of four freshly killed ivory-bills in the possession of a boy in

Gainesville.[12] In the late 1800s, lumbermen in the Gulf Hammock had shot them just to prove they were there. In 1889, Frank M. Chapman, referring to a trip down the last miles of the Suwannee River, just north of the Gulf Hammock area, commented "We go through the best ivory-bill region in the state. . . ."[13]

Although he does not tell us precisely where, Arthur Wayne, a dealer and avid collector, tells us that between 1892 and 1894, he encountered more than 200 ivory-bills in Florida.[14] One can only hope that he did not collect all of them—but I am sure he tried. He and local hunters he hired probably collected nearly that many. In the month of April 1893, he shot thirteen.[15] Wayne also collected a set of three eggs at a Florida ivory-bill nest in 1893.

J. C. Phillips thought that ivory-bills must have existed in high numbers near St. Marks up to about 1905 and that there might be one or two ivory-bills remaining near the mouth of the Aucilla River in north Florida, but added that the area was scheduled for cutting.[16] He also thought there could be some along the Apalachicola River in the early twentieth century.

There is a strong history of ivory-bills from the Gulf coastal swamps associated with Florida rivers from the Apalachicola to the Waccasassa—the entire "Big Bend" area. But, relative to ivory-bill presence elsewhere, we must ask if these were also areas more frequently visited by ornithologists and collectors. Certainly riverine swamp forest habitat was probably nowhere more continuous and abundant.

South and central Florida are areas where a few birds may have persisted well into the twentieth century. Phillips indicated a pair had been at Royal Palm Hammock in 1917, but H. H. Bailey challenged that assertion, stating that he had talked to the warden there and was convinced the birds were pileated woodpeckers, the young of which have a yellow lower bill.[17]

Tanner considered the Big Cypress Swamp of Collier and Hendry Counties in southwest Florida to be second only to the Singer Tract in importance to the future of ivory-bills.[18] Some possible records in the region since Tanner's work are reviewed in Chapter 9.

Other Florida localities for ivory-bills that appear valid, but were missed by Tanner, include (1) an 1890 sight record of an ivory-bill by ornithologist Frank M. Chapman in swamp forest near the entrance to Vista Creek on the Suwannee River in Levy County; (2) an 1891 sight record by Chapman west of the Suwannee River at Fort Fannin, Dixie County; (3) an undocumented 1919 report that ivory-bills occurred in the region of Paradise Key (now Royal Palm Hammock in Everglades National Park); and (4) an incredible "specimen" record that made it into the literature in a 1919 report by Paul Bartsch for the Tortugas.[19] This latter record is almost certainly in error. A. H. Howell suggested that it was "apparently based on an error of a copyist; the original entry in the [US National] museum catalog is without locality."[20] The specimen might have been from Long Key and the original notation misinterpreted; there are two Long Keys, one in the Tortugas and one in the Everglades.[21]

There are many specimen records from Florida for which data do not indicate a specific county locality. Those Florida counties from which specimens are known include Baker, Brevard, Citrus, Dade, DeSoto, Hernando, Hillsborough, Jackson, Jefferson, Lafayette, Lake, Lee, Levy, Manatee, Orange, Osceola, Polk, Taylor, and Volusia.

Georgia

Georgia records for the ivory-bill include sightings from the Savannah River between Augusta and Savannah, from the Satilla and Altamaha Rivers and the Okefenokee Swamp, and from the Chattahoochee River.[22] Two of the three Georgia specimen records include only "Georgia" on the label. The third was from near Waycross, and it is not possible to discern which county.

In the early 1800s, Alexander Wilson was the first to report ivory-bills in the Savannah River Swamp.[23] In 1886, C. B. Prescott shot a male ivory-bill—for which he used the colloquial name of "Snow Kate"—on Miller's Island, just above the mouth of Coleman Lake along the Savannah River, about thirty miles north of Savannah.[24] Eugene Murphey noted that George N. Bailie, an enthusiastic bird student and hunter, had seen an ivory-bill "in a bateau on the Savannah River between Allendale County, South Carolina, and Screven County, Georgia" in September 1907.[25] Murphey added that as the bird flew at treetop level into woods on the Georgia side, Bailie "fired both barrels at it but without even dislodging a feather." (See additional records for the Savannah River Swamp under South Carolina.)

In the late 1800s the artist Rex Brasher had a glimpse "of the Pileated's imperial brother, the Ivory-billed Woodpecker" as he was traveling down the Frederica River in southeast Georgia.[26] He noted that it was "swinging in regulation undulating curves, not fifty feet from the sloop's bow . . . so close [he] could even distinguish its sex—a male." Remembering that ivory-bills usually traveled in pairs, he continued watching and saw the female following about an eighth of a mile behind. He noted "the ladies of the romantic nineties were cautious!" Brasher's observation was about fifteen miles north of Brunswick, Georgia.

The presence of ivory-bills in the great Okefenokee Swamp is mentioned by many authors, but confirmed by few. Sam Mizell is reported to have found three ivory-bills in the region between 1910 and 1915, including one east of the Suwannee Canal, another on Minnie's Island, and a third on Craven's Island.[27] He shot one on Craven's Hammock in 1912, only wounding it. He captured the bird and took it to John M. Hopkins, who in turn gave it to Dan Hebard. Hebard had the bird mounted and kept it on display in his office in Philadelphia until his death in 1941, whereupon the specimen was donated to the Philadelphia Academy of Natural Sciences. Phillips wrote that an ivory-bill was shot in the Okefenokee in 1913, but this report may refer to the 1912 specimen.[28]

A. S. McQueen and Hamp Mizell quote an unidentified resident of the region:

> I have seen these Ivory-bills ever since I was a child, and I did not know they were considered extinct until about 1902 when one was discovered by a scientist. Four have been killed and carried away to museums since 1902. After the natives found out that these birds were found only in the

Okefenokee they stopped killing them, but they seemed to become rarer instead of increasing.[29]

Francis Harper reported hearing an Ivory-bill in the Okefenokee in 1917.[30] In 1920, he wrote of disappearing ivory-bill habitats: "The Minnie Lake Islands, its principal haunt, were reached by lumbering operations about two years ago, and the few remaining birds may have been driven to some other part of the swamp."[31]

I have been able to locate only the Hebard specimen among those mentioned as having come from the Okefenokee Swamp.[32] Certainly some of the many specimens without data could have been from there. Presence of the ivory-bill in the Okefenokee into the 1920s is suggested by McQueen and Mizell, who responded to a statement in *The New International Encyclopedia* that "The finest of the race [of woodpeckers], the great ivory-bill, is now extinct," by saying, "This is an error, for the great Ivory-bill is still in the isolated forest fastness of the Okefenokee, but they are exceedingly rare even in this wonder spot."[33]

Illinois

Audubon noted in his accounts of the ivory-bill that he had seen the bird near the confluence of the Ohio and Mississippi Rivers and as far north as the mouth of the Missouri River. This certainly places it in the vicinity of southern Illinois, if not in the state. Robert Ridgway noted the ivory-billed woodpecker's occurrence in White County, Illinois, in the mid-1800s, stating that he had a "distinct recollection of what he believes to have been" an ivory-bill "some forty miles south of Mount Carmel."[34] This would have been on the Wabash River in eastern Illinois, not far above its confluence with the Ohio River. He also prefaces his remarks by saying that in spite of no recent records, "there can be little doubt . . . that it is still to be found in the heavily timbered counties of the extreme southern portion of the state." Benjamin Gault may have been trying to prove Ridgway's prediction when he reported that he felt "quite certain of hearing its [the ivory-bill's] call note in a swamp near Ullin in Pulaski County in the fall of 1900."[35]

Nearly three quarters of a century later, Ridgway's optimism was viewed by Robert Mengel as "solely a faded recollection of Ridgway's."[36] Firmer evidence of the species having occurred in Illinois was provided by Paul Parmalee, who found a leg bone (tarsometatarsus) of an ivory-bill in an Indian midden in Madison County.[37]

Indiana

R. Haymond, writing in 1869, and Amos W. Butler, in multiple publications between 1885 and 1897, mentioned the ivory-billed woodpecker as having occurred in Franklin and Monroe Counties, in southern Indiana, in the 1800s.[38] In 1897, Butler suggested "there have been reports of its occurrence in [Indiana] . . . in recent years," but he said he could not verify them.[39] Lynds Jones, in 1903, mentioned a specimen of an ivory-billed woodpecker having been taken in Franklin County, Indiana, although I have neither been able to locate it nor to find other details of its existence.[40] Tanner considered the Indiana records as "accidental or as mistaken and unproven identifications," and Mengel, in his *Birds of Kentucky,* considered the Indiana records "anything but convincing."[41]

Kentucky

Mengel provided a good summary of distribution records of ivory-billed woodpeckers in Kentucky.[42] There is no known Kentucky specimen, and no records since the 1870s. Kentucky records include Audubon's report from what is now either Carlisle or Hickman County in bottomlands associated with the Mississippi River in western Kentucky.[43] In 1925, L. Otley Pindar considered evidence for ivory-bills in Fulton County along the Mississippi in southwestern Kentucky.[44] He quoted residents who believed ivory-bills had been common in the area and had been seen as recently as about 1920. Pindar doubted recent reports and that the bird had ever been common in the county. A. W. Schorger and Daniel McKinley found an incredibly detailed eighteenth-century description by Col. William Fleming of a male ivory-billed woodpecker.[45] Fleming had seen two birds, and his servant shot one on March 7, 1780, in Lincoln County, in south central Kentucky, far from the major bottomland forests of the Ohio and Mississippi Rivers.[46]

Louisiana

Harry Church Oberholser described the historical distribution of the ivory-billed woodpecker in Louisiana as being "north to Bayou Mason in West Carroll Parish, Prairie Mer Rouge, and Monroe; west to Ruston and the region about Morgan City; south to La Fourche Parish; and east to Iberville Parish, Saint Joseph, Franklin Parish, and Madison Parish."[47] This encompasses much of the eastern half of the state, with known localities from some of the northernmost and southernmost parishes and a pattern along major rivers that is elaborated on by Tanner.[48] In 1904 there were still ivory-bills in Morehouse, West Carroll, East Carroll, and Madison Parishes—the three northeasternmost parishes in the state. Specimens are known from near Avery Island in Iberia Parish near the coast and from West Carroll and Madison Parishes in the northeast. Other Louisiana specimens lack sufficient data to identify parish of origin. Additional apparently acceptable sight records reported in the literature come from Concordia, East Carroll, Franklin, Lafourche, St. Martin, Tensas, and West Feliciana Parishes—all in the Bayou Macon–Tensas River–Mississippi River bottomland corridor.

In March of 1996, ninety-year-old Stella Christopher of Warrensburg, Missouri, told me of watching ivory-billed woodpeckers in a baldcypress swamp near her home in Red River Parish, in northwestern Louisiana, in 1927 and 1928. Her descriptions convinced me that the birds she watched were indeed ivory-bills. (See also more recent Louisiana reports of ivory-bills in Chapter 9.)

Maryland

Audubon suggested that "now and then an individual . . . may be accidentally found in Maryland," but, in spite of mention by several authors who were probably referring to Audubon's statement, no additional evidence of its existence in Maryland has come to light.[49]

Mississippi

Most Mississippi records of the ivory-bill come from swamps associated with the Sunflower, Yazoo, Tombigbee, Mississippi, and Pascagoula Rivers.[50] John Dennis mentioned ivory-billed woodpecker records from the Pearl River; William H. Turcotte and David Watts suggested they had been present along the Wolf River; and M. G. Vaiden noted that he had known of a pair of ivory-bills in the swamp along the Big Black River in 1908.[51] I have found no other published records or specimens from any of these three areas. Specimens were collected in Harrison and Bolivar Counties in the late 1800s. Julian Corrington saw ivory-bills in the Pascagoula Swamp as recently as December 1921 and noted then that their stronghold was in the remote north and east parts of the swamp and that they were "observed frequently enough . . . on hunting and fishing trips so as to be listed as uncommon rather than rare."[52]

M. G. Vaiden of Rosedale, Mississippi, knew of ivory-bills in Bolivar County up until World War II.[53] Additional counties from which there are literature records of sightings that appear to be acceptable include Clay and Monroe, northeast Mississippi along the Tombigbee River; Hancock and Jackson on the Gulf Coast in bottomlands of the Pearl and Pascagoula Rivers; and Warren, Coahoma, and Yazoo in bottomlands of the Yazoo–Mississippi Delta.

Missouri

Reports of ivory-billed woodpeckers from Fayette in central Missouri and Kansas City in west-central Missouri in the late 1800s and early 1900s were considered by Tanner as accidental or mistaken and unproven identifications.[54] Harry Harris suggested that the species had long ago disappeared from the Kansas City region and noted that it was listed as a former resident of the Missouri River bottomlands on the authority of Judge J. E. Guinotte of Kansas City.[55]

A study skin of an ivory-bill at the Denver Museum of Natural History is labeled as having been collected at Forest Park, near St. Louis and the confluence of the Missouri and Mississippi Rivers in 1886. Otto Widmann noted, "The last record of its capture in Missouri is November, 1895, when Captain Gillespie of the St. Louis police force brought one home from Stoddard Co., and had it mounted by Mr. Frank Schwarz. It was a male and was killed near the Little River on November 8 by a local hunter, named Spradlin, eight miles south of Morely, Scott Co."[56]

Scott County is in southeast Missouri near the confluence of the Ohio and Mississippi Rivers. This specimen has apparently been lost. I visited the Schwarz Studio in St. Louis—now advertised as "America's oldest taxidermy establishment"—and was able to find records of imperial woodpeckers having been mounted there in 1925, but no details of the 1895 ivory-bill.

Although Missouri is rather far north for the species, the AOU (American Ornithologists' Union) Checklist Committee alluded to the possibility of ivory-bills still existing in southern Missouri in 1931.[57] (See Chapter 11 for possible sightings since 1940.)

North Carolina

The most celebrated North Carolina record for the ivory-bill is the bird shot and taken alive by Wilson near Wilmington.[58] A record from Fort Macon was included in Edwin M. Hasbrouck's review and believed to be the northernmost occurrence of the species on the Atlantic coast, but see comments for Maryland.[59] The record was a single sighting reported by Elliott Coues and H. C. Yarrow in 1878.[60]

Ohio

Hasbrouck's distribution map for the species barely reaches into the southwest corner of Ohio, and only archeological specimens provide firm evidence of the species in the state (see Chapter 4).[61] However, in 1903 Lynds Jones wrote, "The evidence of the probable former presence of this regal woodpecker in Ohio is strong. A specimen was taken in Franklin County, Indiana, which borders Ohio."[62]

In his more recent assessment of birds of Ohio, Bruce Peterjohn found no historical evidence of ivory-bills in the state but acknowledged that archaeological evidence suggests that they might have been in the area between five hundred and eight hundred years ago.[63]

Oklahoma

George Sutton summarized Oklahoma records of the distribution of the ivory-billed woodpecker as being in the southeastern part of the state, "northward to the junction of Cimarron and Arkansas Rivers (possibly even to vicinity of Alluwe, along Verdigris River, in Nowata County) and westward to the Blue and Muddy Boggy Rivers in Bryan and Atoka Counties."[64] He considered the last acceptable record for the state to have been in the winter of 1883 to 1884, when W. W. Cooke mentioned the ivory-bill, along with the Carolina chickadee (*Poecile carolinensis*), as indicative of Oklahoma's southern faunal influence.[65] He later noted that the ivory-bill was "not considered by the local hunters as any great rarity."[66] Paul Hahn lists no Oklahoma specimens of ivory-bills, although Sutton mentions a specimen at the Academy of Natural Sciences in Philadelphia.[67] It was supposed to have been obtained from S. W. Woodhouse who visited "Indian Territory" in 1849 to 1850. I examined all specimens at the academy and did not find the specimen in question. Woodhouse, while camped near House Creek in what is now southern Pawnee County, noted on October 13, 1849, that he had seen an ivory-billed woodpecker but that he was not able to get within gunshot of it.[68] Woodhouse also referred to the ivory-bill as "*Picus Eubereas*," an incorrect name that J. S. Tomer and M. J. Brodhead suggested must have been pure guesswork.[69] A 1939 letter to John Baker of the National Audubon Society from Erna Comby of Peco, California, also suggests the possibility of an Oklahoma ivory-bill specimen: "Recently, while we were visiting the State Game Farm at El Reno, Oklahoma, the Supt., Mr. W.A. Gaines . . . showed us a mounted specimen of the Ivory-billed Woodpecker, that is his own property, and claims that there are several of these birds in the Oklahoma Mountains in the southeastern section of the state, around Broken Bow."[70]

South Carolina

Alexander Sprunt and E. B. Chamberlain summarized the former distribution of ivory-billed woodpeckers in South Carolina as extending north to Greenville and Cheraw, east to the lower Santee River and Charleston, south to Hunting Island near Beaufort, and west to Allendale County along the Savannah River. Specimen records come from Chesterfield County and near Charleston (the latter county is not certain).[71] Other counties from which there are apparently acceptable published sight records include Allendale, Barnwell, Beaufort, Berkeley, Charleston, Clarendon, Georgetown, and Greenville.[72]

The ivory-bill was apparently found on several of South Carolina's coastal islands. Walter Hoxie mentioned having collected it on Johnson's, Pritchard's, and Edding Islands.[73] On the basis of a hunter's description, he suggested the presence of ivory-bills on Pritchard's Island as late as 1887, noting that "it was formerly the most visited of all [sea islands] by the Ivory-bill" and that "no one goes there now except an occasional party of deer hunters."[74]

In western South Carolina, T. M. Ashe, a logger who was familiar with ivory-bills in Florida, had seen two or three near Beldoc, Allendale County, South Carolina, in about 1898 when he was cutting the virgin forest.[75] Less than five miles away, in 1907 George N. Bailie shot at and apparently missed an ivory-bill in the Savannah River Swamp between Allendale County, South Carolina, and Screven County, Georgia.[76] Wayne, the prominent South Carolina ornithologist (and logger), who collected many ivory-bills in Florida, lamented that he had never seen the ivory-bill alive in South Carolina, but in 1910 he believed that the bird still could be found in the great swamps that bordered the PeeDee, Santee, and Savannah Rivers, as well as at some inland swamps.[77] (See additional records from the Savannah River Swamp under Georgia.)

In 1934, just after the rediscovery of the birds in Louisiana, George M. Melamphy, a biologist working on a wild turkey project in the Santee River Swamp near Georgetown, reported seeing an ivory-bill about 1930.[78] The National Audubon Society sent Alexander Sprunt, Jr., and Lester L. Walsh to investigate in the spring of 1935. They reported sighting an ivory-bill and hearing possibly two more. Others searched the area, and additional sightings of ivory-bills came in. Sprunt and Walsh leased a portion of the swamp where the birds had been seen and hired Holly Shokes as a warden to guard the area.[79] Shokes later said he saw ivory-bills several times on Wadmacon Island, in the leased area, and that they were almost always in flight. As with the Singer Tract, logging operations on the island began in the late 1930s and continued for about eight years. Most of the big trees were cut and the land was then bulldozed.

It was the opinion of Robin Carter in 1993 that if the ivory-billed survived to recent times anywhere in South Carolina, the Congaree Swamp, southeast of Columbia, in Richland County, offers the best hope.[80] But he said—and I agree—"Don't get your hopes up."

Tennessee

Although it seems likely that ivory-bills occurred in the virgin forests along the Mississippi and Ohio Rivers in Tennessee, there are no credible records of ivory-bills from Tennessee other than those associated with Indian artifacts. Audubon saw two ivory-bills along the Mississippi River north of Fulton, Lauderdale County, in west Tennessee on November 24, 1820, but did not indicate whether they were on the Arkansas or Tennessee side of the river.[81]

Texas

Audubon (or possibly Louis Vieillot; see Chapter 7) was the first to mention ivory-billed woodpeckers in Texas, and he pronounced them as "abundant" in what is now Harris and Ft. Bend Counties.[82] H. E. Dresser reported in 1865 that the ivory-bill was "by no means rare" on the upper Brazos River and "not uncommon" on the Trinity River.[83] An ivory-bill was reported near Russell Creek in Tyler County in about 1880, although by 1885 they were thought to have disappeared from the area.[84] Ivory-bills were also seen near the Neches River before 1910, and one was reported in Angelina County between 1910 and 1915.[85] By 1912, J. K. Strecker considered it as a former inhabitant of east and south Texas ("Harris, Montgomery, Jasper Counties, etc."), but "almost extinct."[86] In his 1926 review, Phillips reported that an ivory-bill had probably been killed near San Antonio in 1900, and that in 1904 a few ivory-bills were heard in east Texas and "at least 1 was shot in Liberty and Harding Counties. . . ."[87] He thought it possible that the birds still existed in the area. Tanner discounted nineteenth-century references to ivory-bills from the San Marco and Guadalupe Rivers and from New Braunfels, Comal County, all in south-central Texas, because of lack of details and information about the observers, although Arthur Allen included these counties as southwestern limits for the species.[88] Harry Church Oberholser and Edgar Kincaid also rejected the New Braunfels record, suggesting that it was a pileated woodpecker.[89]

Texas specimens come from Jasper (eggs), Kaufman, and Liberty Counties.[90] Audubon is said to have collected several specimens of ivory-bills in Harris and Ft. Bend Counties in April or May 1837, although if these specimens still exist, their present locations are unknown.[91] Other counties with apparently acceptable published sight records include Brazoria, Cooke, Dallas, Fannin, Fort Bend, Hardin, Harris, Harrison, Jefferson, Montgomery, Orange, and Tyler.[92]

West Virginia

Although there are no firm records of the ivory-billed woodpecker from West Virginia, there is at least the potential that they occurred there—possibly even following the establishment of settlements by Europeans. George Hall lists the status of the ivory-bill in West Virginia as "hypothetical; extirpated; probably a former casual visitant."[93] He noted that during the brief time that Wilson lived in Shepherdstown, Jefferson

County, West Virginia, he collected an ivory-billed woodpecker some-where between Martinsburg, West Virginia, and Winchester, Virginia. There seems to be no way to know on which side of the present state line the bird was collected. Archeological records of ivory-bills from West Virginia, all of which might have been acquired by Native Americans through trade rather than obtained locally, are discussed in Chapter 4. Karl Haller recounted a conversation with a priest, Father C. Delaux, who told him of shooting and eating a white-billed woodpecker that was larger than an "Indian Hen" in Doddridge County in about 1900.[94]

Appendix 2
Scientific Names of Species
Mentioned in the Text

Plants and animals are included in this appendix under two headings. Within these groupings, the species are listed in alphabetical order using the common names given in the text.

Animals

Acorn woodpecker, *Melanerpes formicivorus*

American crocodile, *Crocodylus acutus*

American redstart, *Setophaga ruticilla*

Bachman's warbler, *Vermivora bachmanii*

Bald eagle, *Haliaeetus leucocephalus*

Barn owl, *Tyto alba*

Barred owl, *Strix varia*

Belted kingfisher, *Ceryle alcyon*

Betsy beetle, *Popilius disjunctus*

Black-and-white warbler, *Mniotilta varia*

Black bear, *Ursus americana*

Black-capped chickadee, *Poecile atricapillus*

Black-throated blue warbler, *Dendroica caerulescens*

Black woodpecker, *Dryocopus martius*

Blue jay, *Cyanocitta cristata*

Brewer's blackbird, *Euphagus cyanocephalus*

California condor, *Gymnogyps californianus*

Cape May warbler, *Dendroica tigrina*

Carolina chickadee, *Poecile carolinensis*

Carolina paroquet (a.k.a. Carolina parakeet), *Conuropsis carolinensis*

Cave swallow, *Petrochelidon fulva*

Common ground-dove, *Columbina passerina*

Common loon, *Gavia immer*

Common raven, *Corvus corax*

Cooper's hawk, *Accipiter cooperii*

Cream-backed woodpecker, *Campephilus leucopogon*

Cuban bullfinch, *Melopyrrha nigra*

Cuban crow, *Corvus nasicus*

Cuban emerald, *Chlorstilbon ricordii*

Cuban gnatcatcher, *Polioptila lembeyei*

Cuban grassquit, *Tiaris canora*

Cuban kestrel, *Falco sparverius sparveroides*

Cuban ivory-billed woodpecker, *Campephilus principalis bairdi*

Cuban parrot, *Amazona leucocephala*

Cuban pewee, *Contopus caribaeus*

Cuban solitaire, *Myadestes elisabeth*

Cuban tody, *Todus multicolor*

Cuban trogan, *Priotelus temnurus*

Cuban vireo, *Vireo gundlachii*

Dodo, *Raphus cuculatus*

Downy woodpecker, *Picoides pubescens*

Eastern bluebird, *Sialia sialis*

Eastern cougar, *Felis concolor*

Florida panther, *Felis concolor coryi*

Galapagos penguin, *Spheniscus mendiculus*

Giant armadillo, *Priodontes maximus*

Gray rat snake, *Elaphe obsoleta spiloides*

Gray squirrel, *Sciurus carolinensis*

Gray wolf, *Canis lupus*

Great auk, *Pinguinus impennis*

Great horned owl, *Bubo virginianus*

Great slaty woodpecker, *Mulleripicus pulverulentus*

Hairy woodpecker, *Picoides villosus*

Hardwood-stump borer, *Mallodon dasystomus* (formerly *Stenodontes dasystomus*)

Honeybee, *Apis melifera*

Horned guan, *Oreophasis derbianus*

Horned passalus, *Popilius disjunctus* (a.k.a. Betsy beetle)

House sparrow, *Passer domesticus*

Imperial woodpecker, *Campephilus imperialis*

Ivory-billed woodpecker, *Campephilus principalis*

Labrador duck, *Camptorhynchus labradorius*

Long-horned beetle, *Parandra polita*

Louisiana paraquet, (a.k.a. Carolina parakeet), *Conuropsis carolinensis*

Magnolia warbler, *Dendroica magnolia*

Mallard, *Anas platyrhynchos*

Mantled howler monkey, *Alouatta palliata*

Mourning dove, *Zenaida macroura*

Northern bobwhite, *Colinus virginianus*

Northern cardinal, *Cardinalis cardinalis*

Northern parula, *Parula americana*

Northern pintail, *Anas acuta*

Nuthatch, *Sitta* sp.

Opposum, *Didelphis virginianus*

Orange-crowned warbler, *Vermivora celata*

Oriente warbler, *Teretistris fornsi*

Pacific Ridley sea turtle, *Lepidochelys olivacea*

Palm warbler, *Dendroica palmarum*

Pampas deer, *Ozotoceros bezoarticus bezoarticus*

Peregrine falcon, *Falco peregrinus*

Pileated woodpecker, *Dryocopus pileatus*

Prairie warbler, *Dendroica discolor*

Raccoon, *Procyon lotor*

Red-bellied woodpecker, *Melanerpes carolinus*

Red-cockaded woodpecker, *Picoides borealis*

Red-headed woodpecker, *Melanerpes erythrocephalus*

Red-necked pigeon (a.k.a. = Scaly-naped pigeon), *Columba squamosa*

Red-shouldered hawk, *Buteo lineatus*

Red-tailed hawk, *Buteo jamaicensis*

Red-winged blackbird, *Agelaius phoeniceus*

Red wolf, *Canis rufus*

Rusty blackbird, *Euphagus carolinus*

Scarlet tanager, *Piranga olivacea*

South American ivory-billed woodpecker, *Campephilus* spp.

Southern pine beetle, *Dendroctonus frontalis*

Spanish lynx, *Lynx lynx oardinus*

Turkey vulture, *Cathartes aura*

Vicuna, *Vicugna vicugna*

White-tailed deer, *Odocoileus virginianus*

Wild hog, *Sus scrofa*

Wild turkey, *Meleagris gallopavo*

Wood duck, *Aix sponsa*

Yellow-bellied sapsucker, *Sphyrapicus varius*

Yellow-rumped warbler, *Dendroica coronata*

Yellow-shafted flicker, *Colaptes auratus*

Yellow-tailed woolly monkey, *Lagothrix flavicauda*

Zapata sparrow, *Torreornis inexpectata*

Plants

American elm, *Ulmus americana*

Ash, *Fraxinus* sp.

Baldcypress, *Taxodium distichum*

Basswood, *Tilia americana*

Black gum, *Nyssa sylvatica*

Black oak, *Quercus velutina*

Black willow, *Salix nigra*

Boxelder, *Acer negundo*

Burr oak, *Quercus macrocarpa*

Cabbage palm, *Sabal palmetto*

Cottonwood, *Populus deltoides*

Cuban pine, *Pinus cubensis*

Elm, *Ulmus* sp.

Grape, *Vitis* sp.

Hackberry, *Celtis* sp.

Hickory, *Carya* sp.

Honey locust, *Gleditsia triacanthos*

Laurel oak, *Quercus laurifolia*

Live oak, *Quercus virginiana*

Loblolly pine, *Pinus taeda*

Longleaf pine, *Pinus palustris*

Nuttall oak, *Quercus nuttalli*

Oak, *Quercus* sp.

Overcup oak, *Quercus lyrata*

Palmetto, *Sabal palmetto*

Pecan, *Carya illinoensis*

Persimmon, *Diospyros virginiana*

Pine, *Pinus* spp.

Poison ivy, *Toxicodendron radicans*

Redbay, *Persea borbonia*

Red maple, *Acer rubrum*

Red mulberry, *Mores rubra*

Sawgrass, *Cladium jamaicense*

Southern magnolia, *Magnolia grandiflora*

Southern red oak, *Quercus falcata*

Soybean, *Glycine max*

Spruce pine, *Pinus glabra*

Sugarberry, *Celtis laevigata*

Sugarcane, *Saccharum*

Sweetbay, *Magnolia virginiana*

Sweetgum, *Liquidambar styraciflua*

Switchcane, *Arundinaria gigantea*

Sycamore, *Platanus occidentalis*

Tupelo gum, *Nyssa aquatica*

Water oak, *Quercus nigra*

White elm, *Ulmus americana* (a.k.a. American elm)

Willow oak, *Quercus phellos*

Notes

Introduction

1. R. K. Selander, "Sexual Dimorphism and Differential Niche Utilization in Birds," *Condor* 68 (1966): 113–51. 2. J. T. Tanner, *The Ivory-Billed Woodpecker, National Audubon Society, Research Report No. 1* (New York: National Audubon Society, 1942). 3. B. Romans, *A Natural History of East and West Florida* (Gainesville: University of Florida Press, 1771/1962).

1. Behavior and Ecology: How It Might Still Live

1. B. Christy, "The Vanishing Ivory-Bill," *Audubon Magazine* 45, no. 2 (1943): 99–102. 2. J. A. Jackson, "Red-Cockaded Woodpecker *Picoides borealis*," in *The Birds of North America, No. 85*, ed. A. Poole and F. Gill (Philadelphia: Academy of Natural Sciences; Washington, DC: American Ornithologists' Union, 1994), 1–20; W. D. Koenig and R. L. Mumme, *Population Ecology of the Cooperatively Breeding Acorn Woodpecker* (Princeton, NJ: Princeton University Press, 1987). 3. W. E. D. Scott, *The Story of a Bird Lover* (New York: Macmillan, 1903). 4. G. H. Lowery, "The Ivory-Billed Woodpecker in Louisiana," *Proceedings of the Louisiana Academy of Science* 2 (1935): 84–86. 5. J. V. Dennis, "A Last Remnant of Ivory-Billed Woodpeckers in Cuba," *Auk* 65 (1948): 497–07. 6. J. T. Tanner, *The Ivory-Billed Woodpecker* (Research Report No. 1) (New York: National Audubon Society, 1942). 7. J. A. Jackson and H. R. Ouellet, "Downy Woodpecker *Picoides pubescens,*" in *The Birds of North America, No. 613,* ed. A. Poole and F. Gill (Philadelphia: Birds of North America), 1–32; Jackson, "Red-Cockaded Woodpecker." 8. J. J. Audubon and J. B. Chevalier, *The Birds of America. Volume 4* (New York: Dover Publications, 1840–1844/1967),

269

215–16. 9. J. J. Audubon, *Journal of John James Audubon Made During His Trip to New Orleans in 1820–1821*, ed. H. Corning (Cambridge, MA: Business Historical Society, 1929), 39. 10. Lowery, "The Ivory-Billed Woodpecker in Louisiana." 11. A. A. Allen, "Ivory-Billed Woodpecker," in *Life Histories of North American Woodpeckers*, ed. A. C. Bent (Washington, DC: US National Museum Bulletin, 1939), 1–12. 12. H. Winkler, D. A. Christie, and D. Nurney, *Woodpeckers. A Guide to the Woodpeckers of the World* (Boston: Houghton Mifflin, 1995), 393. 13. Allen, "Ivory-Billed Woodpecker"; A. A. Allen, "Hunting with a Microphone the Voices of Vanishing Birds"; *National Geographic Magazine* 71, no. 6 (1937): 696–723; A. A. Allen and P. P. Kellogg, "Recent Observations on the Ivory-Billed Woodpecker," *Auk* 54 (1937): 164–84. 14. Audubon and Chevalier, *The Birds of America*. 15. F. M. Chapman, *Handbook of Birds of Eastern North America*, 6th ed. (New York: D. Appleton), 229. 16. Allen and Kellogg, "Recent Observations on the Ivory-Billed." 17. R. D. Hoyt, "Nesting of the Ivory-Billed Woodpecker in Florida," *Warbler* (2nd ser.) 1 (1905): 52–55. 18. S. C. Graham, "The Ivory-billed Woodpecker," *Forest & Stream* 72 (1909): 892. 19. Lowery, *The Ivory-Billed Woodpecker in Louisiana*, 85–86. 20. T. G. Pearson, "Handsome Flickers and a Rare Cousin," in *The Book of Birds, Vol. 2*, ed. G. Grosvenor and A. Wetmore (Washington, DC: National Geographic Society, 1939), 64–73. 21. Audubon and Chevalier, *The Birds of America*. 22. E. A. McIlhenny, cited in C. Bendire, "Life Histories of North American Birds, From the Parrots to the Grackles, with Special Reference to Their Breeding Habits and Eggs," *US National Museum Special Bulletin* 3, 43. 23. G. E. Beyer, "The Ivory-Billed Woodpecker in Louisiana," *Auk* 17 (1900): 97–99. 24. J. T. Tanner, *The Ivory-Billed Woodpecker* (Research Report No. 1) (New York: National Audubon Society, 1942), 57. 25. Allen and Kellogg, "Recent Observations on the Ivory-Billed." 26. J. J. Audubon, *Journal of John James Audubon Made During His Trip to New Orleans in 1820–1821*, ed. H. Corning (Cambridge, MA: Business Historical Society, 1929), 77–78. 27. J. V. Dennis, "The Ivory-Billed Woodpecker (*Campephilus principalis*)," *Avicultural Magazine* 85 (1979): 75–84. 28. McIlhenny, in Bendire, *Life Histories of North American Birds*; M. Thompson, "A Red-Headed Family," *Hoosier Naturalist* 2, no. 5 (1887): 73–76. 29. McIlhenny, in Bendire, *Life Histories of North American Birds*. 30. H. Winkler, D. A. Christie, and D. Nurney, *Woodpeckers. A Guide to the Woodpeckers of the World* (Boston: Houghton Mifflin, 1995); L. L. Short, *Woodpeckers of the World* (Greenville: Delaware Museum of Natural History, 1982). 31. Tanner, *The Ivory-Billed Woodpecker*, 62. 32. Ibid, 57. 33. Letter from H. Stoddard to J. Tanner, September 29, 1939. 34. Allen, "Ivory-Billed Woodpecker," 2. 35. Ibid., 8. 36. J. A. Jackson, H. R. Ouellet, and B. J. S. Jackson, "Hairy Woodpecker (*Picoides villosus*)," in *The Birds of North America, No. 702*, ed. A. Poole and F. Gill (Philadelphia: Birds of North America, 2002), 1–32. 37. J. J. Audubon and J. B. Chevalier, *The Birds of America. Volume 4* (New York: Dover, 1840–1844/1967), 217. 38. F. E. L. Beal, *Food of the Woodpeckers of the United States* (Washington, DC: US Department of Agriculture, Biological Survey Bulletin No. 37, 1911). 39. C. Cottam and P. Knappen, "Food of Some Uncommon North American Birds," *Auk* 56 (1939): 138–69. 40. R. W. Doane, E. C. Van Dyke, W. J. Chamberlin, and H. E. Burke, *Forest Insects* (New York: McGraw-Hill, 1936). 41. R. Headstrom, *The Beetles of North America* (New York: A. S. Barnes, 1977). 42. G. R. Lamb, *The Ivory-Billed Woodpecker in Cuba* (New York: Research Report No. 1, Pan-American Section, International Committee for Bird Preservation, 1957); J. A. Jackson, "Ivory-Billed Woodpecker (*Campephilus principalis*)," in *The Birds of North America, No. 711*, ed. A. Poole and F. Gill (Philadelphia: Birds of North America, 2002), 1–28. 43. Ibid.; interview with Arelardo García, May 1987; Allen and Kellogg, "Recent Observations on the Ivory-Billed";

F. García, *Las aves de Cuba. Subespecies endemicas. Tomo II* (Havana, Cuba: Gente Nueva, 1987). 44. Allen and Kellogg, "Recent Observations on the Ivory-Billed"; J. T. Tanner, *The Ivory-Billed Woodpecker. Research Report No. 1* (New York: National Audubon Society, 1942). 45. Tanner, *The Ivory-Billed Woodpecker*; W. E. D. Scott, "On Birds Observed in Sumpter, Levy, and Hillsboro' Counties, Florida," *Bulletin of the Nuttall Ornithological Club* 6 (1881): 14–21. 46. Hoyt, "Nesting of the Ivory-Billed Woodpecker." 47. Ibid. 48. J. A. Jackson, "How to Determine the Status of a Woodpecker Nest," *Living Bird* 15 (1976): 205–21. 49. W. E. D. Scott, "Notes on the Birds of the Caloosahatchie Region of Florida," *Auk* 9 (1892): 209–18. 50. Tanner, *The Ivory-Billed Woodpecker*. 51. McIlhenny, in Bendire, "Life Histories of North American Birds." 52. Beyer, "The Ivory-Billed Woodpecker in Louisiana." 53. H. B. Bailey, "Memoranda of a Collection of Eggs from Georgia," *Bulletin of the Nuttall Ornithological Club* 8 (1883): 37–43. 54. M. Thompson, "A Red-Headed Family," *Library Magazine* 13, no. 30 (1885): 289–93. 55. J. J. Audubon and J. B. Chevalier, *The Birds of America. Volume 4* (New York: Dover, 1840–1844/1967). 56. Tanner, *The Ivory-Billed Woodpecker*. 57. Ibid. 58. Jackson, "Red-Cockaded Woodpecker." 59. J. S. Y. Hoyt, "Preliminary Notes on the Development of Nestling Pileated Woodpeckers," *Auk* 61 (1944): 376–84; E. L. Bull and J. A. Jackson, "Pileated Woodpecker (*Dryocopus pileatus*)," in *The Birds of North America, No. 148*, ed. A. Poole and F. Gill (Philadelphia: Academy of Natural Sciences; Washington, DC: American Ornithologists' Union, 1995), 1–24. 60. L. Kilham, "Courtship and the Pair-Bond of Pileated Woodpeckers," *Auk* 96 (1979): 587–94. 61. S. Cramp, *Handbook of the Birds of Europe, the Middle East, and North Africa. Vol. 4* (Oxford: Oxford University Press, 1985). 62. Scott, *The Story of a Bird Lover*, 269. 63. Jackson, "How to Determine the Status." 64. A. A. Allen and P. P. Kellogg, "Recent Observations on the Ivory-Billed," 179. 65. Tanner, *The Ivory-Billed Woodpecker*, 62. 66. Beyer, "The Ivory-Billed Woodpecker in Louisiana." 67. Tanner, *The Ivory-Billed Woodpecker*. 68. Audubon and Chevalier, *The Birds of America*; Scott, *The Story of a Bird Lover*; W. E. D. Scott, "Supplementary Notes from the Gulf Coast of Florida, with a Description of a New Species of Marsh Wren," *Auk* 5 (1888): 183–88. 69. Scott, "Supplementary Notes from the Gulf Coast." 70. Tanner, *The Ivory-Billed Woodpecker*. 71. McIlhenny, in Bendire, *Life Histories of North American Birds*. 72. Audubon and Chevalier, *The Birds of America*, 216. 73. H. H. Bailey, *The Birds of Florida* (Baltimore: Williams and Wilkins, 1925). 74. Scott, *The Story of a Bird Lover*. 75. Beyer, "The Ivory-Billed Woodpecker in Louisiana"; Tanner, *The Ivory-Billed Woodpecker*. 76. Beyer, "The Ivory-Billed Woodpecker in Louisiana," 98. 77. Ibid. 78. Short, *Woodpeckers of the World*. 79. Beyer, "The Ivory-Billed Woodpecker in Louisiana." 80. Tanner, *The Ivory-Billed Woodpecker*. 81. Audubon and Chevalier, *The Birds of America*, p. 217. 82. Yale University specimen # 4633, purchased from General Biological Supply House, Chicago, Illinois, April 26, 1939; a female; data suggest the bird was collected January 15, 1890, at Kissimmee, Florida. 83. Audubon and Chevalier, *The Birds of America*, 217. 84. Tanner, *The Ivory-Billed Woodpecker*. 85. Ibid. 86. Bull and Jackson, "Pileated Woodpecker." 87. J. A. Jackson and O. H. Dakin, "An Encounter Between a Nesting Barn Owl and a Gray Rat Snake," *Raptor Research* 16 (1982): 60–61. 88. Tanner, *The Ivory-Billed Woodpecker*. 89. Short, *Woodpeckers of the World*, 451. 90. Tanner, *The Ivory-Billed Woodpecker*. 91. Short, *Woodpeckers of the World*. 92. Ibid., 451.

2. The Land of the Ivory-Bill

1. A. Wilson, *American Ornithology. Vol. 4* (Philadelphia: Bradford and Inskeep, 1811). 2. J. T. Tanner, Mimeographed text of a statement made to the National Association of Audubon

Societies, October 1939. 3. P. Brodkorb, "Catalogue of Fossil Birds: Part 4 (Columbiformes through Piciformes)," *Bulletin of the Florida State Museum,* 15, no. 4 (1971): 163–66. 4. Short considers the ivory-bill and imperial woodpeckers to be members of a superspecies. L. L. Short, "The Habits and Relationships of the Magellanic Woodpecker," *Wilson Bulletin* 82 (1970): 115–29. 5. D. Garcilasco, *La Florida del Inca* (Madrid, Spain: Privately published, 1723), cited in E. L. Fundaburk and M. D. Foreman, *Sun Circles and Human Hands* (Tuscaloosa: University of Alabama Press, 2001), 35. 6. R. F. Johnston and R. K. Selander, "House Sparrow: Rapid Evolution of Races in North America," *Science* 144 (1964): 548–50. 7. E. M. Hasbrouck, "The Present Status of the Ivory-Billed Woodpecker (*Campephilus principalis*)," *Auk* 8 (1891): 174–86, 184. 8. A. A. Allen, "Ivory-Billed Woodpecker," in *Life Histories of North American Woodpeckers* (Washington, DC: US National Museum, 1939), 1–12; J. T. Tanner, *The Ivory-Billed Woodpecker. Research Report No. 1* (New York: National Audubon Society, 1942). 9. A. W. Schorger, "An Early Record and Description of the Ivory-Billed Woodpecker in Kentucky," *Wilson Bulletin* 61 (1949): 235; D. McKinley, "Early Record for the Ivory-Billed Woodpecker in Kentucky," *Wilson Bulletin* 70 (1958): 380–81. 10. Tanner, *The Ivory-Billed Woodpecker,* 14. 11. J. A. Jackson, "Tree Surfaces as Foraging Substrates for Insectivorous Birds," in *The Role of Insectivorous Birds in Forest Ecosystems,* ed. J. G. Dickson, R. N. Conner, R. R. Fleet, J. C. Kroll, and J. A. Jackson (New York: Academic Press, 1979), 69–93. 12. Tanner, *The Ivory-Billed Woodpecker.* 13. Except for J. A. Jackson, "The Southeastern Pine Forest Ecosystem and Its Birds: Past, Present, and Future," in *Bird Conservation 3* (Madison: University of Wisconsin Press, 1988), 119–59. 14. S. C. Graham, "The Ivory-Billed Woodpecker," *Forest & Stream* 72 (1909): 892. 15. A. H. Howell, *Florida Bird Life* (Tallahassee: Florida Department of Game and Fresh Water Fish, 1932), 314. 16. Wilson, *American Ornithology.* 17. Ibid. 18. H. C. Oberholser and E. B. Kincaid, Jr., *The Bird Life of Texas. Vol. 1* (Austin: University of Texas Press, 1974), 528. 19. H. Nehrling, "List of Birds Observed at Houston, Harris Co., Texas, and in the Counties Montgomery, Galveston, and Fort Bend," *Bulletin of the Nuttall Ornithological Club* 7 (1882): 166–75. 20. J. A. Putnam, *Management of Bottomland Hardwoods* (New Orleans, LA: US Department of Agriculture, Forest Service, Southern Forest Experiment Station, 1951).

3. Recognition and Causes of Decline

1. C. C. Abbott, *The Birds About Us* (Philadelphia: J.B. Lippincott, 1895). 2. E. M. Hasbrouck, "The Present Status of the Ivory-Billed Woodpecker (*Campephilus principalis*)," *Auk* 8 (1891): 174–86. 3. P. Laurent,"Notes on the Birds of Levy Co., Florida," *Ornithologist and Oölogist* 12 (1887): 157–59; P. Laurent, "Bird Notes from a Florida Porch," *Bird-Lore* 8 (1906): 67. 4. J. C. Phillips, *An Attempt to List the Extinct and Vanishing Birds of the Western Hemisphere* (Copenhagen, Denmark: Verhandlungen VI Internationalen Ornithologen-Kongresses Kopenhagen, 1926), 512–13. 5. J. A. Jackson, "Red-Cockaded Woodpecker *Picoides borealis,*" in *The Birds of North America, No. 85,* ed. A. Poole and F. Gill (Philadelphia: Academy of Natural Sciences; Washington, DC: American Ornithologists' Union, 1994), 1–20. 6. R. G. Lillard, *The Great Forest* (New York: Alfred A. Knopf, 1947). 7. T. C. Croker Jr., "The Longleaf Pine," *Journal of Forest History* 23 (1979): 32–43. 8. J. V. Dennis, *The Great Cypress Swamps* (Baton Rouge: Louisiana State University Press, 1988); J. A. Jackson, "The Southeastern Pine Forest Ecosystem and Its Birds: Past, Present, and Future," in *Bird Conservation 3* (Madison: University of Wisconsin Press, 1988), 119–59. 9. H. C. Oberholser and E. B. Kincaid, Jr., *The Bird Life of Texas. Vol. 1* (Austin: University of Texas Press,

1974). 10. Letter from M. G. Vaiden to James Bond, May 20, 1963, Academy of Natural Sciences, Philadelphia. 11. Ibid. 12. M. L. Comeaux, *Atchafalaya Swamp Life. Settlement and Folk Occupations* (Baton Rouge: School of Geoscience, Louisiana State University, 1972). 13. R. C. Bryant, *Logging: The Principles and General Methods of Operation in the United States,* 2nd ed. (New York: J. Wiley and Sons, 1923). 14. J. V. Dennis, *The Great Cypress Swamps* (Baton Rouge: Louisiana State University Press, 1988). 15. P. O. MacDonald, W. E. Frayer, and J. K. Clauser, *Documentation, Chronology, and Future Projections of Bottomland Hardwood Habitat Loss in the Lower Mississippi Alluvial Plain. Vol. I: Basic Report* (Washington, DC: Ecological Services, Fish and Wildlife Service, US Department of the Interior, 1979). 16. A. H. Howell, *Florida Bird Life* (Tallahassee: Florida Department of Game and Fresh Water Fish, 1932), 314; A. A. Allen and P. P. Kellogg, "Recent Observations on the Ivory-Billed Woodpecker," *Auk* 54 (1937): 164–84. 17. C. N. Elliott, "Feathers of the Okefenokee," *American Forests* 38 (1932): 202–6, 253. 18. E. M. Coulter, *A Short History of Georgia* (Chapel Hill: University of North Carolina Press, 1933). 19. V. H. Cahalane, A. Leopold, W. L. Finley, and C. Cottam, "Report of the Committee on Bird Protection, 1939," *Auk* 57 (1940): 279–91; A. Sprunt and E. B. Chamberlain, *South Carolina Bird Life* (Columbia: University of South Carolina Press, 1970). 20. J. T. Tanner, *The Ivory-Billed Woodpecker. Research Report No. 1* (New York: National Audubon Society, 1942); D. Eckelberry, "Search for the Rare Ivorybill," in *Discovery. Great Moments in the Lives of Outstanding Naturalists,* ed. J. K. Terres (Philadelphia: J.B. Lippincott, 1961), 195–207. 21. Allen and Kellogg, "Recent Observations of the Ivory-Billed," 184. 22. J. C. Greenway, Jr., *Extinct and Vanishing Birds of the World* (New York: Special Publication No. 13, American Committee for International Wild Life Protection, 1958). 23. P. Laurent, "My Ivory-Billed Woodpeckers," *Oologist* 34 (1917): 65–67. 24. T. D. Burleigh, *Georgia Birds* (Norman: University of Oklahoma Press, 1958), 366. 25. A. T. Wayne, "Notes on the Birds of the Wacissa and Aucilla River Regions of Florida," *Auk* 12 (1895): 362–67. 26. P. F. Balboa, *Las aves de Cuba* (Havana, Cuba: Cultural Sociedad Anónima, 1941); J. V. Dennis, "A Last Remnant of Ivory-Billed Woodpeckers in Cuba," *Auk* 65 (1948): 497–507. 27. J. J. Audubon and J. B. Chevalier, *The Birds of America. Vol. 4* (New York: Dover Publications, 1840–1844/1967), 216. 28. Wayne, "Notes on the Birds of the Wacissa." 29. Advertisement of Chas. K. Worthen, facing title page of *Nidologist* 3, no. 4 (1896). 30. H. R. Taylor, [Editorial comment], *Nidologist* 2, no. 10 (1895): 146. 31. H. R. Taylor, [Editorial comment], *Nidologist* 3, no. 7 (1896): 80. 32. L. F. Kiff and D. J. Hough, *Inventory of Bird Egg Collections of North America, 1985* (Norman, OK: American Ornithologists' Union and Oklahoma Biological Survey, 1985). 33. Anonymous [a "Committee of twenty-five prominent American oologists"], *The American Oologists' Exchange Price List of North American Birds' Eggs* (Lacon, IL: R. Magoon Barnes, 1922). 34. S. G. Kohlstedt, "Henry A. Ward: The Merchant Naturalist and American Museum Development," *Journal of the Society for the Bibliography of Natural History* 9 (1980): 647–61. 35. Anonymous, [Editorial comment], *Nidologist* 3 (1895): 12. 36. P. Steinhart, "A Common Possession," in *The National Audubon Society. Speaking for nature. A Century of Conservation,* ed. L. Line (New York: Hugh Lauter Levin Associates, 1999), 42–59. 37. W. Dutcher et al., [Report of AOU Committee for Protection of North American Birds], *Auk* 22 (1905): 110–12. 38. P. Hahn, *Where Is that Vanished Bird?* (Toronto, Ontario, Canada: Royal Ontario Museum, University of Toronto Press, 1963). 39. M. Stouffer, *Marty Stouffer's Wild America* (New York: Times Books, 1988), 101–02. 40. Phillips, *An Attempt to List.* 41. H. H. Bailey, "The Ivory-Billed

Woodpecker in Florida," *Oologist* 44 (1927): 18, 20. 42. W. D. Miller, W. G. Van Name, and D. Quinn, *A Crisis in Conservation. Serious Danger of Extinction of Many North American Birds* (New York: Emergency Conservation Committee, 1929). 43. W. Cuppy, *How to Become Extinct* (New York: Farrar & Rinehart, 1941). 44. B. Christy, "The Vanishing Ivory-Bill," *Audubon Magazine* 45, no. 2 (1943): 99–102. 45. T. G. Pearson, "Handsome Flickers and a Rare Cousin," in *The Book of Birds, Vol. 2*, ed. G. Grosvenor and A. Wetmore (Washington, DC: National Geographic Society, 1939), 64–73.

4. In the Time of the Tribes

1. A. M. Bailey, "Ivory-Billed Woodpecker's Beak in an Indian Grave in Colorado," *Condor* 41 (1939): 164. 2. P. W. Parmalee, "Additional Noteworthy Records of Birds from Archaeological Sites," *Wilson Bulletin* 79 (1967): 155–62. 3. H. Van der Schalie and P. W. Parmalee, "Animal Remains from the Etowah Site, Mound C Bartow County, Georgia," *Florida Anthropologist* 13, nos. 2–3 (1960): 37–54. 4. P. W. Parmalee, "Remains of Rare and Extinct Birds from Illinois Indian Sites," *Auk* 75 (1958): 169–76. 5. P. W. Parmalee, "Vertebrate Remains from an Historic Archaeological Site in Rock Island County, Illinois," *Transactions of the Illinois State Academy of Sciences* 57 (1964): 167–74. 6. J. M. O'Shea, G. D. Schrimper, and J. K. Ludwickson, "Ivory-Billed Woodpeckers at the Big Village of the Omaha," *Plains Anthropologist* 27, no. 97 (1982): 245–48. 7. B. C. Yates, "Faunas from House 5 at the Vinson Site," *Bulletin of the Texas Archeological Society* 64 (1993): 187–225. 8. A. Wetmore, "Evidence of the Former Occurrence of the Ivory-Billed Woodpecker in Ohio," *Wilson Bulletin* 55 (1943): 55. 9. J. L. Murphy and J. Farrand, Jr., "Prehistoric Occurrence of the Ivory-Billed Woodpecker (*Campephilus principalis*), Muskingum County, Ohio," *Ohio Journal of Science* 79 (1979): 22–23. 10. J. E. Guilday, "Biological and Archeological Analysis of Bones from a 17th Century Indian Village (46 PU 31), Putnam County, West Virginia," *West Virginia Geological and Economic Survey, Report of Archeological Investigation No. 4* (1971), 1–64; P. Parmalee, personal communication, cited in G. A. Hall, *West Virginia Birds* (Pittsburgh, PA: Carnegie Museum of Natural History, 1983). 11. Parmalee, personal communication. 12. J. Clayton, [Letter to the Royal Society giving an account of the soil and other things seen in Virginia], *Philosophical Transactions of the Royal Society, London* 17 (1693): 978–99. 13. H. McBurney, *Mark Catesby's Natural History of America. The Watercolors from the Royal Library Windsor Castle* (London: Merrell Holberton, 1997), 46, and quoted from M. Catesby, *Natural History of Carolina, Florida and the Bahama Islands, Vol. 1* (London: C. Marsh, 1731). 14. M. R. Harrington, "Sacred Bundles of the Sac and Fox Indians," *University of Pennsylvania Museum of Anthropolology Publications* 4, no. 2 (1914): 123–262, esp. 169, 226. 15. Ibid. 16. A. Wilson, *American Ornithology, Volume 4* (Philadelphia: Bradford and Inskeep, 1811), 20–26, pl. 29. 17. J. J. Audubon, *Ornithological Biography. Volume 1* (Edinburgh, Scotland: A. and C. Black, 1831), 343. 18. E. D. Crabb, *The Woodpeckers of Oklahoma,* (Norman: University of Oklahoma Press, Publications of the University of Oklahoma Biological Survey 2(3), 1930). 19. M. R. Blaine, *The Ioway Indians* (Norman: University of Oklahoma Press, 1979). 20. Data with the pipe stems at the Milwaukee Public Museum indicate that one (MPM# 30133) was considered a sacred pipe of the Aruhwa or female Buffalo gens. The second (MPM# 30135) was said to have been used as a substitute for the first one and was also a sacred pipe of the Aruhwa or female Buffalo gens. The third (MPM# 30536) had been obtained from a member of the "Buffalo Gens." The fourth (MPM# 30137) is labeled as a pipe stem of the "Pigeon gens." 21. A. Skinner, "Ethnology of the

Ioway Indians," *Bulletin of the Public Museum of Milwaukee* 5, no. 4 (1926): 181–354, esp. 224. 22. A. C. Fletcher and F. LaFlesche, *The Omaha Tribe. The Twenty-Seventh Annual Report of the Bureau of American Ethnology to the Secretary of the Smithsonian Institution 1905–1906* (Washington, DC: US Government Printing Office, 1911). 23. From Charles Willoughby, as cited in E. L. Fundaburk and M. D. F. Foreman, *Sun Circles and Human Hands* (Luvurne, AL: Emma Lila Fundaburk, 1957), 57. 24. Personal interview with Carla Dove, August 2001. 25. S. Vestal, *Warpath* (Lincoln: University of Nebraska Press, 1984), 186. 26. E. Ingersoll, *Birds in Legend Fable and Folklore* (New York: Longman's Green, 1923), 235. 27. T. Page, *The Civilization of the American Indians* (New York: Crescent Books, 1979). 28. J. Michelet, *The Bird* (Edinburgh, Scotland: T. Nelson and Sons, 1869), 226–27. 29. W. J. Hoxie, "A Seminole Vocabulary," *Atlantic Slope Naturalist* 1 (1903): 66–67. 30. O'Shea, Schrimper, and Ludwickson, "Ivory-Billed Woodpeckers." 31. L. J. P. Vieillot, *Histoire naturelle des oiseaux de l'Amerique septentrionale. Volume 2* (Paris: Chez Desray, 1807). 32. W. Z. Parks, cited in A. M. Bailey, "Ivory-Billed Woodpecker's Beak in an Indian Grave in Colorado," *Condor* 41 (1939): 164. 33. See pl. 46 in Fundaburk and Foreman, *Sun Circles and Human Hands*. 34. J. P. Brain and P. Phillips, *Shell Gorgets. Styles of the Late Prehistoric and Protohistoric Southeast* (Cambridge, MA: Peabody Museum Press, Peabody Museum of Archaeology and Ethnology, Harvard University, 1996). 35. Ibid. 36. Ibid.; G. E. Lankford, *Native American Legends. Southeastern Legends: Tales from the Natchez, Caddo, Biloxi, Chickasaw, and Other Nations* (Little Rock, AR: August House, 1987), 73; W. H. Holmes, *Art in Shell of the Ancient Americans* (Washington, DC: Bureau of American Ethnology, 1883). 37. R. J. Wheeler, "Metal Crested Woodpeckers: Artifacts of the Terminal Glades Complex," *Florida Anthropologist* 50 (1997): 67–81; R. J. Wheeler, *Treasure of the Calusa. The Johnson/Willcox Collection from Mound Key, Florida* (Tallahassee: Monographs in Florida Archaeology, 2000). 38. Wheeler, "Metal Crested Woodpeckers." 39. Anonymous, *A Brief Description of a Creek Accession and Ascension Ceremony Conducted During the Harvest Busk Held at Pine Arbor Tribal Town at Blountstown, Florida, November 1995* (2002), http//:www.freenet.thl.fl.us/Museum/culture/mekko.htm 40. Ibid.

5. The Discoveries of Early Naturalists

1. J. Lawson, *A New Voyage to Carolina* (Chapel Hill: University of North Carolina Press, 1709/1967). 2. There is some debate about the year of Catesby's birth because of changes in the calendar at the time. *See* G. F. Frick and R. P. Stearns, *Mark Catesby, the Colonial Audubon* (Urbana: University of Illinois Press, 1961). 3. E. G. Allen, "The History of American Ornithology Before Audubon," *Transactions of the American Philosophical Society* 41, pt. 3 (1951): 387–591, and frontispiece and title page. 4. M. Catesby, *Natural History of Carolina, Florida and the Bahama Islands, Vol. 1* (London: C. Marsh, 1731). 5. T. D. Burleigh, *Georgia Birds* (Norman: University of Oklahoma Press, 1958). 6. A. Feducia, ed., *Catesby's Birds of Colonial America* (Chapel Hill: University of North Carolina Press, 1985), 88. 7. A. W. Schorger, "An Early Record and Description of the Ivory-Billed Woodpecker in Kentucky," *Wilson Bulletin* 61 (1949): 235; D. McKinley, "Early Record for the Ivory-Billed Woodpecker in Kentucky," *Wilson Bulletin* 70 (1958): 380–81. 8. Schorger, "An Early Record and Description." 9. Ibid. 10. Bartram, *Travels Through North and South Carolina, Georgia, East & West Florida, the Cherokee Country, the Extensive Territories of the Muscogulges, or Creek Confederacy, and the Country of the Chactaws: Containing an Account of the Soil and Natural Productions of Those Regions, Together with Their Observations on the Manners of Indians," in The Travels of William Bartram*, ed., F. Harper (Athens: University of Georgia Press,

1998), xxxvii–lxii, 1–332. 11. E. Coues, "Fasti ornithologiae.—No. I. Bartram's 'Travels,'" *Proceedings of the Academy of Natural Sciences, Philadelphia* 27 (1875): 338–58. 12. Bartram, "Travels Through North & South Carolina,"56. 13. C. L. Remington, "[John Abbot's] Notes on My Life," *Lepidopterists' News* 2 (March 1948): 28–30; E. E. Murphey, "Observations on the Bird Life of the Middle Savannah Valley 1890–1937," *Contributions from the Charleston Museum* 9 (1937): i–vii, 1–61. 14. Allen, "The History of American Ornithology." 15. Murphey, "Observations on the Bird Life of the Middle Savannah." 16. C. Hunter, *The Life and Letters of Alexander Wilson* (Philadelphia: American Philosophical Society, 1983), 86. 17. Ibid., 397. 18. W. Faxon, "John Abbot's Drawings of the Birds of Georgia," *Auk* 13 (1896): 204–15. 19. V. Rogers-Price, "Introduction and Commentary," in *John Abbot's Birds of Georgia*, ed. G. M. Arthur (Savannah, GA: Beehive Press, 1997), vii–xlii. 20. Ibid. 21. Allen, "The History of American Ornithology"; P. H. Oesher, "Louis Jean Pierre Vieillot (1748–1831)," *Auk* 65 (1948): 568–76. 22. I have pieced together a brief introduction to Alexander Wilson from several sources to provide some insight into the background, character, and motivation of the man who would become known as the "father of American ornithology" and who provided key early insights into the behavior of ivory-billed woodpeckers. References used include Allen, "The History of American Ornithology Before Audubon"; R. Cantwell, *Alexander Wilson Naturalist and Pioneer* (Philadelphia: J. B. Lippincott, 1961); Hunter, *The Life and Letters of Alexander Wilson;* G. Ord, *Sketch of the Life of Alexander Wilson* (Philadelphia: Harrison Hall). 23. Cantwell, *Alexander Wilson Naturalist and Pioneer,* 115. 24. Hunter, *The Life and Letters of Alexander Wilson,* 203. 25. A. Wilson, *American Ornithology* (Philadelphia: Bradford and Inskeep, 1811). 26. Hunter, *The Life and Letters of Alexander Wilson,* 378. 27. W. Hoxie, "Probable Occurences of the Ivory-Billed Woodpecker on Pritchard's Island, South Carolina," *Ornithologist & Oologist* 12 (1887): 122. 28. A. Wilson and C. L. Bonaparte, *American Ornithology; or the Natural History of the Birds of the United States* (Edinburgh, Scotland: Constable, 1831), 132–38. 29. A. J. Tyler, *I Who Should Command All* (New York: G.P. Putnam's Sons, 1942). 30. H. Corning, *Journal of John James Audubon Made During His Trip to New Orleans in 1820–1821* (Cambridge, MA: Business Historical Society, 1929), 33. 31. Ibid., 36. 32. Ibid., 39. 33. Ibid., 77–78. 34. Ibid., 82–83. 35. Ibid., 184. 36. J. J. Audubon and J. B. Chevalier, *The Birds of America. Vol. 4* (New York: Dover, 1840–1844/1967). 37. The original edition of Audubon's *The Birds of America* was published as four volumes between 1827 and 1838. It was known as the *Double Elephant Folio,* a reference to the large size of the paper on which it was printed. By definition, paper that is "double elephant" is 40 inches by 26.5 inches; Audubon's engraver used paper that was 39.5 inches by 29.5 inches—apparently close enough to earn the name. The ivory-billed woodpecker was included in Volume 4 of the *Double Elephant Folio.* J. J. Audubon, *The Birds of America. Vol. 4* (London: Robert Havell, 1984). 38. Audubon and Chevalier, *The Birds of America,* 222. 39. A. Ford, *James John Audubon: A Biography* (New York: Abbeville Press).

6. Arthur Allen

1. M. Harwood, "The Lab: From Hatching to Fledging," in *Cornell Laboratory of Ornithology. Annual Report 1986–87* (Ithaca, NY: Cornell Laboratory of Ornithology, 1967), 4–13. 2. E. W. Teale, "Arthur A. Allen," *Audubon Magazine* 46 (1943): 84–89. 3. A. H. Howell, *Florida Bird Life* (Tallahassee: Florida Department of Game and Fresh Water Fish, in cooperation with Bureau of Biological Survey, US Department of Agriculture, 1932), 314. 4. A. A. Allen, "Vacationing with Birds," *Bird-Lore* 26

(1924): 208–13, 211. 5. Photo in the Arthur Allen collection in the Department of Manuscripts & University Archives, Cornell University Libraries, Ithaca, NY. 6. Letter from A. A. Allen to M. P. Tindel [Tindall; misspelled in original], June 19, 1924; in the Arthur Allen collection in the Department of Manuscripts & University Archives, Cornell University Libraries, Ithaca, NY. 7. A. R. Bird, "Ivory-Bill Is Still King!" *American Forests* 38 (1932): 634–35, 667. 8. J. T. Tanner, *The Ivory-Billed Woodpecker. Research Report No. 1* (New York: National Audubon Society, 1942), 6. 9. Field Museum of Natural History, Chicago, Illinois, specimen #139,362. 10. University of Florida female specimen #18988, card #39882, and male specimen #18992, card #39881; accession #2013. There is additional confusion regarding these specimens because the museum's ivory-billed woodpecker specimens that now bear the numbers 18987, 18988, and 18992 had been unlabeled, mounted specimens, and the correct number assignment is unknown. The similar secondary feather wear of specimens 18992 and 18988 suggest that these two were the pair collected by Hancock. 11. Tanner, *The Ivory-Billed Woodpecker,* 6. 12. Harwood, "The Lab." 13. T. S. Palmer, "The Forty-Ninth Stated Meeting of the American Ornithologists' Union October 19–22, 1931," *Auk* 49 (1932): 52–64. 14. C. N. Elliott, " Feathers of the Okefenokee," *American Forests* 38 (1932): 202–6, 253. 15. Bird, "Ivory-Bill Is Still King!", 634. 16. G. E. Beyer, "The Ivory-Billed Woodpecker in Louisiana," *Auk* 17 (1900): 97–99. 17. G. H. Lowery, Jr., *Louisiana Birds* (Baton Rouge: Louisiana State University Press, 1974). 18. T. G. Pearson, "Protection of the Ivory-Billed Woodpecker," *Bird-Lore* 34 (1932): 300–01; E. G. Holt, "Report of Ernest G. Holt, Department of Sanctuaries," *Bird-Lore* 34 (1932): 441–45. 19. A. A. Allen, "Hunting with a Microphone the Voices of Vanishing Birds," *National Geographic Magazine* 71 (1937): 696–723. 20. Letter from A. A. Allen to J. J. Kuhn, Tallulah, Louisiana, Department of Manuscripts and University Archives, Cornell University Libraries, Ithaca, NY. 21. M. Stouffer, *Marty Stouffer's Wild America* (New York: Times Books, 1988).

7. James Tanner

1. J. T. Tanner, "A Melanistic Black-Capped Chickadee," *Auk* 51 (1934): 240. 2. A. A. Allen, "Hunting with a Microphone the Voices of Vanishing Birds," *National Geographic Magazine* 71 (1937): 696–723. 3. J. T. Tanner, "Sound Recordings for a Natural History Museum" (master's thesis, Cornell University, Ithaca, NY, 1936). 4. T. S. Palmer, "The Fifty-Second Stated Meeting of the American Ornithologists' Union," *Auk* 52 (1935): 53–63. 5. J. T. Tanner, *The Ivory-Billed Woodpecker. Research Report No. 1* (New York: National Audubon Society, 1942), iii. 6. Letter from A. Allen to J. Tanner, January 14, 1937; letter provided courtesy of N. Tanner. 7. Letter from J. Baker to A. Allen; in the Arthur Allen collection in the Department of Manuscripts & University Archives, Cornell University Libraries, Ithaca, NY. 8. Letter from A. Allen to J. Tanner. 9. J. T. Tanner, "Observations in Madison Parish, Louisiana," *Auk* 56 (1939): 90. 10. J. T. Tanner, "Bird-Lore's Thirty-Eighth Christmas Census. (Singer Tract)," *Bird-Lore* 40 (1938): 54. 11. L. E. Hicks, "The Fifty-Sixth Stated Meeting of the American Ornithologists' Union," *Auk* 56 (1939): 112–24. 12. J. T. Tanner,"The Last Wilderness of the Mississippi Bottomlands," *Louisiana Conservation Review* 9, no. 2 (1940): 13–16, 49. 13. Letter from J. Tanner to A. Allen, January 14, 1942; letter from A. Allen to J. Tanner, January 7, 1942. Letters provided courtesy of N. Tanner. 14. Letter from R. T. Peterson to J. Tanner, June 15, 1942; letter provided courtesy of N. Tanner; R. T. Peterson, "My Greatest Birding Moment," *Audubon* (Suppl.) 90, no. 2 (1988): 64. 15. Letter of

R. T. Peterson to J. Tanner. 16. J. T. Tanner, "The Ivory-Billed Woodpecker," *Texas Game and Fish* 14, no. 7 (1956): 15, 30–31. 17. J. T. Tanner, "Distribution of Tree Species in Louisiana Bottomland Forests," *Castanea* 51 (1986): 168–74.

8. The Struggle for the Singer Tract

1. R. H. Pough, *On the Present Condition of the Tensas River Forests of Madison Parish, Louisiana and the Status of the Ivory-Billed Woodpecker in this Area as of January, 1944,* Unpublished Report to the executive director, National Audubon Society, NY, 1944. 2. Ibid. 3. E. A. McIlhenny, "The Passing of the Ivory-Billed Woodpecker," *Auk* 58 (1941): 582–84. 4. A. R. Bird, "Ivory-Bill Is Still King!" *American Forests* 38 (1932): 634–35, 667. 5. E. G. Holt, "Ivory-Billed Woodpeckers," *Bird-Lore* 34 (1932): 441–45. 6. G. H. Lowery, "The Ivory-Billed Woodpecker in Louisiana," *Proceedings of the Louisiana Academy of Sciences* 2 (1935): 84–86. 7. G. H. Lowery, Jr., *Louisiana birds,* 3rd ed. (Baton Rouge: Louisiana State University Press, 1974); Lowery, *The Ivory-Billed Woodpecker in Louisiana;* J. S. Campbell et al., "Bird-Lore's Thirty-Fourth Christmas Census (Tallulah, La.)," *Bird-Lore* 36 (1934): 55. 8. Letter from A. Allen to A. Leopold, May 4, 1936, and four-page draft manuscript titled "Proposal for a Conservation Inventory of Threatened Species" by A. Leopold, Department of Manuscripts & University Archives, Cornell University Libraries, Ithaca, NY. 9. Letter of G. H. Bick to J. Tanner, August 8, 1941. Bick was a biologist working for the Louisiana Department of Conservation. 10. McIlhenny, "The Passing of the Ivory-Billed Woodpecker." 11. Anonymous, "Urge National Park in Tensas Parish," *Louisiana Conservation Review* 10 (1941): 9; C. A. Cottam et al., "Report of the Committee on Bird Protection, 1941," *Auk* 59 (1942): 286–99. 12. V. H. Cahalane et al., "Report of the Committee on Bird Protection, 1939," *Auk* 57 (1940): 279–91, esp. 291; V. H. Cahalane et al., "Report of the Committee on Bird Protection, 1940," *Auk* 58 (1941): 292–98. 13. Letter from G. H. Bick to J. Tanner, August 8, 1941. Letter provided courtesy of N. Tanner. 14. J. Tanner, field notes. From Department of Manuscripts and University Archives, Cornell University Libraries, Ithaca, NY. 15. Letter from A. Allen to J. Tanner, March 25, 1942. Department of Manuscripts and University Archives, Cornell University Libraries, Ithaca, NY. 16. J. H. Baker, "The Director Reports to You," *Audubon Magazine* 44 (1942): 367–76. 17. Ibid.; A. N. Pack, [Editorial note], *Nature Magazine* 35 (1942): 382. 18. H. B. Chase, Jr., "Preserve Endangered," *Nature Magazine* 35 (1942): 382. 19. S. 78-468 (1942), endnote. 20. R. Edge, "The Singer Tract and the Ivory-Billed Woodpecker," in *Conservation for Victory* (New York: Emergency Conservation Committee, 1943), 22. 21. J. H. Baker, Mimeograph of text of a speech titled, "Statement by John H. Baker, Executive Director, National Audubon Society, with Regard to Establishment of Wildlife Refuge in Louisiana in an Effort, Among Other Things, to Preserve America's Rarest Bird," presented at the Convention of Outdoor Writers Association of America in Columbus, OH, February 22, 1944. Copy courtesy of N. Tanner. 22. Pack, Editorial note. 23. Edge, "The Singer Tract and the Ivory-Billed Woodpecker." 24. Ibid. 25. D. R. Mandell, "Timber Production War Project," in *Encyclopedia of American Forest and Conservation History. Vol. 2,* ed. R. C. Davis (New York: Macmillan, 1983), 647. 26. Baker, "The Director Reports to You." 27. Baker, "Statement by John H. Baker." 28. Ibid. 29. Anonymous, "Rare Bird Vanishing," *Pennsylvania Game News* 14, no. 11 (1944): 30. 30. Baker, "Statement by John Baker." 31. R. Pough, Unpublished report to J. Baker, National Audubon Society, February 16, 1944; copy courtesy of N. Tanner. 32. Ibid., 9. 33. Letter from J. Tanner to R. H. Pough, February 4, 1946; copy courtesy of

N. Tanner. 34. T. Barbour, *That Vanishing Eden. A Naturalist's Florida* (Boston: Little, Brown, 1944), 72. 35. D. Eckelberry, "Search for the Rare Ivorybill," in *Discovery. Great Moments in the Lives of Outstanding Naturalists*, ed. J. K. Terres (Philadelphia: J.B. Lippincott, 1961), 195–207; T. Angell, "The Ivory-Bill's Last Stand," *Living Bird* 21, no. 4 (2002): 11–16. 36. Letter from A. MacMurray to J. Tanner, January 8, 1949; courtesy of N. Tanner. 37. Ibid. 38. J. V. Dennis, "The Ivory-Billed Woodpecker (*Campephilus principalis*)," *Avicultural Magazine* 85 (1979): 75–84. 39. H. Ossa, *They Saved Our Birds* (New York: Hippocrene Books, 1982), 205–06. 40. Baker, " Statement by John Baker." 41. Letter from J. Tanner to H. T. Stone, refuge manager, Tensas River National Wildlife Refuge, April 2, 1986; copy courtesy of N. Tanner. 42. Copy of tentative program, "Dedication Ceremony, Tensas River National Wildlife Refuge Facilities Office/Visitor Center, Shop Complex," courtesy of N. Tanner.

9. In the Footsteps of Others

1. J. Zickefoose, "Ivory-Billed Woodpecker," *Bird Watcher's Digest* 21, no. 5 (1999): 28–43; C. Cokinos, *Hope Is the Thing with Feathers* (New York: Jeremy P. Tarcher/Putnam, 2000); J. A. Jackson, "The History of Ivory-Billed Woodpeckers in Mississippi," *Mississippi Kite* 18 (1988): 3–10; J. A. Jackson, "Ivory-Billed Woodpecker *Campephilus principalis principalis*," in *Rare and Endangered Biota of Florida. Birds*, ed. J. A. Rodgers and H. W. Kale, II (Gainesville: University of Florida Press, 1996), 103–12; J. A. Jackson, "Ivory-Billed Woodpecker *Campephilus principalis*," in *The Birds of North America, No. 711*, ed. A. Poole and F. Gill (Philadelphia: The Birds of North America, 2003), 1–28. 2. J. A. Jackson, "An Evaluation of Aerial Survey Techniques for Red-Cockaded Woodpeckers," *Journal of Wildlife Management* 49 (1985): 1083–88. 3. J. T. Tanner, *The Ivory-Billed Woodpecker. Research Report No. 1* (New York: National Audubon Society, 1942), 21. 4. Ibid. 5. Letter from D. James to J. A. Jackson, June 4, 1987. 6. Ibid.; J. Neal, University of Arkansas, personal communication, July 1988. 7. W. Eastman, "Ten Year Search for the Ivory-Billed Woodpecker," *Atlantic Naturalist* 13 (1958): 216–28; D. H. Crompton, "My Search for the Ivory-Billed Woodpecker in Florida," *Massachusetts Audubon Society Bulletin* 34, no. 6 (1950): 235–37. 8. J. H. Baker, "News of Wildlife and Conservation; Ivory-Bills Now Have Sanctuary," *Audubon Magazine* 52 (1950): 391–92. 9. J. V. Dennis, "The Ivory-Bill Flies Still," *Audubon Magazine* 69, no. 6 (1967): 38–44. 10. Ibid.; J. V. Dennis, " The Ivory-Billed Woodpecker (*Campephilus principalis*)," *Avicultural Magazine* 85 (1979): 75–84. 11. R. H. Pough, "Bird Protection in the United States," *Bulletin of the International Committee for Bird Preservation* VI (1952): 224–29. 12. Letter from D. Lee to J. A. Jackson, August 28, 1990, and unpublished manuscript by D. Lee. 13. H. N. Agey and G. M. Heinzmann, "The Ivory-Billed Woodpecker Found in Central Florida," *Florida Naturalist* 44 (1971): 46–47, 64. 14. Dennis, "The Ivory-Bill Flies Still"; J. V. Dennis, *The Great Cypress Swamps* (Baton Rouge: Louisiana State University Press, 1988), 42. 15. Letter from A. Wetmore to G. M. Heinzmann, April 7, 1968. 16. Dennis, "The Ivory-Bill Flies Still." 17. Letter from W. B. Robertson to J. A. Jackson, October 20, 1986. 18. Jackson, "Ivory-Billed Woodpecker." 19. J. K. Terres, [Article on ivory-billed woodpeckers], *Linnaean Society of New York Newsletter* (December 1986); J. K. Terres, Copy of a manuscript dated January 19, 1987, correcting the locality for Terres's ivory-billed woodpecker sighting described in 1986 article (1987). This may have been published in a subsequent issue of the *Linnaean Society of New York Newsletter*. Copy provided by T. Engstrom. 20. Dennis, "The Ivory-Billed Woodpecker."

21. Unpublished manuscript in US Fish and Wildlife Service files, Atlanta, GA. (The files included a cover letter from D. G. Garratt to Regional Director, US Fish and Wildlife Service, dated July 2, 1985. The manuscript that was included was titled, "Possible Sightings of an Ivory-Billed Woodpecker *(Campephilus principalis)* in Jonathan Dickinson State Park South Central Florida.") 22. G. A. Patterson and W. B. Robertson, Jr., "Distribution and Habitat of the Red-Cockaded Woodpecker in Big Cypress National Preserve," *South Florida Research Center Report* T-613 (1981): i–vi, 1–137. 23. E.g., D. S. Maehr et al., "Fates of Wild Hogs Released into Occupied Florida Panther Home Ranges," *Florida Field Naturalist* 17 (1989): 43–45. 24. Letter from W. B. Robertson to J. A. Jackson. 25. W. Brewster and F. M. Chapman, "Notes on the Birds of the Lower Suwanee River," *Auk* 8 (1891): 125–38. 26. W. Brewster, "With the Birds on a Florida River," *Bulletin of the Nuttall Ornithological Club* 6 (1881): 38–44. 27. L. Renner, "Wekiva's Allure Has Drawn Generations of Humans," *Orlando Sentinel,* December 6, 1987, p. A9. 28. Fide Jim Lewis, US Fish and Wildlife Service, personal communication, August 1988. 29. Eastman, "Ten Year Search." 30. P. Sykes, Report to Director, Patuxent Wildlife Research Center, November 27, 1967. 31. C. N. Elliott, "Feathers of the Okefenokee," *American Forests* 38 (1932): 202–06, 253. 32. T. D. Burleigh, *Georgia Birds* (Norman: University of Oklahoma Press, 1958). 33. Ibid., 365. 34. G. Heinrich and C. Welch, "A Natural History and Current Conservation Projects to Save the Ivory-Billed Woodpecker, *Campephilus principalis principalis," AAZPA [American Association of Zoological Parks and Aquariums] Regional Conference Proceedings* 1983 (1983): 224–29; Dennis, "The Ivory-Billed Woodpecker." 35. Dennis, "The Ivory-Billed Woodpecker." 36. A. A. Allen and P. P. Kellogg, "Recent Observations on the Ivory-Billed Woodpecker," *Auk* 54 (1937): 164–84; J. T. Tanner, *The Ivory-Billed Woodpecker. Research Report No. 1* (New York: National Audubon Society, 1942). 37. H. Bick, "Ivory-Billed Woodpecker and Wild Turkeys in Louisiana," *Auk* 59 (1942): 431–32; R. T. Peterson, "My Greatest Birding Moment," *Audubon* (Suppl.) 90, no. 2 (1988): 64. 38. Peterson, "My Greatest Birding Moment." 39. Letter from L. C. Binford to J. A. Jackson, November 29, 1991. 40. Heinrich and Welch, "A Natural History." 41. J. T. Tanner, "Distribution of Tree Species in Louisiana Bottomland Forests," *Castanea* 51 (1986): 168–74. 42. G. H. Lowery, Jr., *Louisiana birds,* 3rd ed. (Baton Rouge: Louisiana State University Press, 1974). 43. Dennis, "The Ivory-Billed Woodpecker." 44. Ibid.; R. B. Hamilton, "Central Southern Region," *American Birds* 29 (1975): 700–05. 45. Hamilton, "Central Southern Region." 46. J. V. Dennis, *The Great Cypress Swamps* (Baton Rouge: Louisiana State University Press, 1988). 47. W. E. Sommer, "Atchafalaya Basin Levee Construction" (master's thesis, Tulane University, New Orleans, Louisiana, 1966). 48. Dennis, *The Great Cypress Swamps,* 109. 49. S. G. Thigpen, Sr., *Pearl River: Highway to Glory Land* (Kingsport, TN: Kingsport Press, 1965). 50. W. Stolzenburg, "Swan Song of the Ivory-Bill," *Nature Conservancy* 52, no. 3 (2002): 38–47. 51. J. W. Fitzpatrick, *Acoustic Search for the Ivory-Billed Woodpecker in the Pearl River Wildlife Management Area and the Bogue Chitto National Wildlife Refuge, Louisiana (Winter 2002),* Final Report, May 31, 2002, Cornell Laboratory of Ornithology. 52. Ibid.; T. Angell, "The Ivory-Bill's Last Stand," *Living Bird* 21, no. 4 (2002): 11–16. 53. J. Rosen, "The Ghost Bird," *New Yorker* (May 14, 2001): 61–67; J. J. Williams, "Ivory-Billed Dreams, Ivory-Billed Reality," *Birding* 33 (2001): 514–27. 54. G. E. Moore, "Ivory-Bills Again," *Bluebird* 21, no. 5 (1954): 2. 55. D. G. Schueler, *Preserving the Pascagoula* (Jackson: University Press of Mississippi, 1980). 56. J. D. Corrington, "The Winter Birds of the Biloxi, Mississippi, Region," *Auk* 39 (1922): 545. 57. Schueler, *Preserving the Pascagoula,* 99 58. The name of this individual is "George M. Melamphy," not "George E. Lamprey," as given in T. G. Pearson, C. S. Brimley, and

H. H. Brimley, *Birds of North Carolina,* rev. by D. L. Wray and H. T. Davis (Raleigh: North Carolina Department of Agriculture, State Museum, 1959); Letter from J. Tanner to J. A. Jackson, September 11, 1989. 59. A. Sprunt and E. B. Chamberlain, *South Carolina Bird Life* (Columbia: University of South Carolina Press, 1970). 60. C. Cottam et al., "Report to the American Ornithologists' Union by the Committee on Bird Protection, 1961," *Auk* 79 (1962): 463–78. 61. Letters from P. Sykes to Director, Patuxent Wildlife Research Center, September 12, 13, and 14, 1970. 62. Anonymous, "Signal from the Wild," *Time* 97 (March 29, 1971): 49. 63. G. Able and J. Moran, *Paddling South Carolina* (Columbia, SC: Palmetto Byways Press, 1986). 64. E. B. Chamberlain, *Rare & Endangered Birds of the Southern National Forests* (Atlanta, GA: USDA Forest Service, 1974). 65. Able and Moran, *Paddling South Carolina.* 66. Tanner, *The Ivory-Billed Woodpecker,* 23. 67. H. C. Oberholser and E. B. Kincaid, Jr., *The Bird Life of Texas. Vol. 1* (Austin: University of Texas Press, 1974); C. E. Shackelford, "A Compilation of Published Records of the Ivory-Billed Woodpecker in Texas: Voucher Specimens Versus Sight Records," *Bulletin of the Texas Ornithological Society* 31 (1998): 34–41. 68. Eastman, "Ten Year Search." 69. J. Tanner, personal communication, January 30, 1986. 70. G. Reynard, personal communication, April 4, 1988; C. Cottam et al., "Report to the American Ornithologists' Union," 174; Shackelford, "A Compilation of Published Records. 71. Eastman, "Ten Year Search." 72. J. W. Hardy, "A Tape Recording of a Possible Ivory-Billed Woodpecker Call," *American Birds* 29 (1975): 647–51. 73. Tanner, personal communication; P. W. Sykes, Jr., "Report on the Search for the Ivory-Billed Woodpecker *(Campephilus principalis principalis)* in Eastern Texas, January 1968," Undated, unpublished report to US Department of the Interior, Fish and Wildlife Service, Bureau of Sport Fisheries and Wildlife, Delray Beach, Florida. 74. G. M. Sutton, personal communication, October 16, 1974. 75. Tanner, personal communication. 76. P. Street, *Wildlife Preservation* (New York: Henry Regnery, 1971), 105. 77. Letter from G. Reynard to J. A. Jackson, April 4, 1988. 78. Letter from H. Owens to J. A. Jackson, October 21, 1987. 79. Dennis, "The Ivory-Billed Woodpecker"; Letter from R. B. Mounsey to J. A. Jackson, March 25, 1977. 80. Dennis, "The Ivory-Billed Woodpecker." 81. G. Reynard and O. Garrido, *Bird Songs in Cuba* (Ithaca, NY: Cornell Laboratory of Ornithology, 1989). 82. Letter from J. Tanner to J. A. Jackson, September 11, 1989. 83. Dennis, *The Great Cypress Swamps,* 42.

10. Cuba

1. C. B. Cory, "The Birds of the West Indies, Including the Bahama Islands, the Greater and the Lesser Antilles, Excepting the Islands of Tobago and Trinidad," *Auk* 3 (1886): 337–81. 2. J. Cassin, "Notes on the Picidae," *Proceedings of the Academy of Natural Sciences, Philadelphia, Pennsylvania* 1863 (1864): 322–28. 3. S. F. Baird, T. M. Brewer, and R. R. Ridgway, *A History of North American Birds* (Boston: Little, Brown, 1874). 4. R. de la Sagra, "Énumération des espèces zoologiques et botaniques de l'île de Cuba utiles à acclimater dans d'autres régions analogues du globe," *Bulletin de la Société Zoologique d'Acclimation* [Paris] 6 (1859): 169–84 5. J. Gundlach, "Revista y catalogo de les aves Cubanas," in *Repertorio Fisico Natural de la Isla de Cuba, Vol. 1,* ed. F. Poey (Havana, Cuba: Imprenta del Gobierno y Capitania General, 1866); J. Gundlach, "Neue beitrëge zur ornithologie Cubas," *Journal für Ornithologie* 22 (1874): 113–66; J. Gundlach, *Contribucion a la ornitología Cubana* (Havana, Cuba: Imprenta La Antilla, 1876); J. Gundlach, *Ornitología Cubana* (Havana, Cuba: Imprenta La Habana, 1893). 6. N. J. Collar et al., *Threatened Birds of the Americas: The ICBP Red Data Book* (Cambridge: International Council for Bird Preservation, 1992); O. H.

Garrido and A. Kirkconnell, *Field Guide to the Birds of Cuba* (Ithaca, NY: Cornell University Press, 2000). 7. W. Stone, "Barbour's 'Birds of Cuba,'" *Auk* 40 (1923): 548–49. 8. P. F. Balboa, *Las aves de Cuba* (Havana, Cuba: Cultural Sociedad Anónima, 1941); J. V. Dennis, "A Last Remnant of Ivory-Billed Woodpeckers in Cuba," *Auk* 65 (1918): 497–507. 9. J. Bond, *Check-List of Birds of the West Indies* (Philadelphia: Academy of Natural Sciences, 1940); Dennis, "A Last Remnant of Ivory-Billed Woodpeckers." 10. L. S. Rowe and P. de Alba, *Documentary Material on Nature Protection and Wild Life Preservation in Latin America. Vol. I, Part I. Fauna* (Washington, DC: Pan American Union, Division of Agricultural Cooperation, 1940). 11. T. Barbour, "The Birds of Cuba," *Memoirs of the Nuttall Ornithological Club* 6 (1923): 91. 12. A. Moreno, *Necesidad de proteger nuestras aves. Memorias de la Sociedad Cubana de Historia Natural* (Havana, Cuba: Sociedad Cubana de Historia Natural, 1938). 13. Bond, *Check-List of Birds.* 14. T. Barbour, "Cuban Ornithology," *Nuttall Ornithological Club* 9 (1943): 86. 15. J. V. Dennis, "Davis Crompton and the Cuban Ivory-Billed Woodpecker," *BirdWatcher's Digest* 9, no. 4 (1987): 18–25. 16. J. Bond, *Check-List of Birds of the West Indies, 4th ed.* (Philadelphia: Academy of Natural Sciences, 1956), 103. 17. G. R. Lamb, *The Ivory-Billed Woodpecker in Cuba* (New York: Pan-American Section, International Committee for Bird Preservation, 1957). 18. Letter from B. W. Read to W. Eastman, September 18, 1956. 19. J. Bond, *Thirteenth Supplement to the Checklist of Birds of the West Indies (1956)* (Philadelphia: Academy of Natural Sciences, 1968); F. Garcia, *Las aves de Cuba. Subspecies endemicas. Vol. II* (Havana, Cuba: Gente Nueva, 1987). 20. K. Curry-Lindahl, *Let Them Live* (New York: William Morrow, 1972), 220. 21. R. L. Norton, "West Indies Region," *American Birds* 35 (1981): 338. 22. Anonymous, [Advertisement for Bird Bonanzas, inc. [sic], Miami, Florida], *American Birds* 35 (1981): 986. 23. O. H. Garrido, "Cuban Endangered Birds," in *Neotropical Ornithology*, ed. P. A. Buckley et al.. *Ornithological Monographs* 36 (1985): 992–99; 24. A. M. Shull, "Endangered and Threatened Wildlife and Plants: Review of the Status of the Ivory-Billed Woodpecker," *Federal Register* 50 (1985): 14123–24. 25. L. L. Short and J. F. M. Horne, "The Ivory-Bill Still Lives," *Natural History* 95, no. 7 (1986): 26, 28. 26. L. L. Short, "Last Chance for the Ivory-Bill," *Natural History* 94, no. 8 (1985): 66–68. 27. A. R. Estrada and G. Alayón García, "Reporte de expedición: Búsqueda de Carpintero Real," *Volante Migratorio* 6 (1986): 15. 28. Short and Horne, "The Ivory-Bill Still Lives"; Estrada and Alayón García, "Reporte de expedicíon"; A. R. Estrada and G. Alayón García, "La existencia del Carpintero Real o Pico de Marfil en Cuba, es realidad, no es sueño," *Volante Migratorio* 7 (1986): 25–27; G. B. Reynard, "The Ivory-Billed Woodpecker in Cuba," in *Proceedings of the Third Southeast Nongame Endangered Wildlife Symposium*, ed. R. R. Odom, K. A. Riddleberger, and J. C. Ozier (Athens: Georgia Department of Natural Resources, Game and Fish Division, 1987), 8–10. 29. L. L. Short and J. F. M. Horne, "I Saw It!" *International Wildlife* 17 (March/April 1987): 22–23. 30. Estrada and Alayón García, "La existencia del Carpintero Real o Pico de Marfil en Cuba." 31. *See also* J. A. Jackson, "Will-o'-the-Wisp," *Living Bird Quarterly* 10, no. 1 (1991): 29–32. 32. J. A. Jackson, "Habitat Conditions in the Vicinity of Ivory-Billed Woodpecker Sightings in Eastern Cuba," *El Pitirre* 3, no. 3 (1990): 7. 33. J. Gundlach, "Revista y catalogo de les aves Cubanas"; J. Gundlach, Neue beitrëge zur ornithologie Cubas." 34. J. McNeely, *Ivory-Billed Woodpecker Expedition Report,* Unpublished report; M. Lammertink, "Search for Ivory-Billed Woodpecker in Cuba," *Dutch Birding* 14 (1992): 170–73. 35. Lammertink, "Search for Ivory-Billed." 36. Ibid.; M. Lammertink, "No More Hope for the Ivory-Billed Woodpecker *Campephilus principalis,*" *Cotinga* 3 (1995): 45–47; M. Lammertink and A. R. Estrada, "Status of the Ivory-Billed Woodpecker *Campephilus principalis* in Cuba:

Almost Certainly Extinct," *Bird Conservation International* 5 (1995): 53–59. Reynard, "The Ivory-Billed Woodpecker in Cuba." 38. Lammertink, "Search for Ivory-Billed." 39. Lammertink and Estrada, "Status of the Ivory-Billed." 40. Lammertink, "Search for Ivory-Billed"; Lammertink, "No More Hope." 41. Garrido and Kirkconnell, *Field Guide to the Birds of Cuba.* 42. Lammertink, "Search for Ivory-Billed." 43. Lammertink, "No More Hope."

11. An American Icon

1. J. J. Audubon and J. B. Chevalier, *The Birds of America. Vol. 4* (New York: Dover, 1840–1844/1967). 2. J. A. Jackson, "The Truth Is Out There," *Birder's World* 16, no. 3 (2002): 40–47. 3. R. Edge, "The Singer Tract and the Ivory-Billed Woodpecker," in *Conservation for Victory* (New York: Emergency Conservation Committee, Annual Report, 1942, 1943). 4. M. B. Davidson, *The Original Water-Color Paintings by John James Audubon for* The Birds of North America (New York: American Heritage, 1966). 5. W. Faulkner, "The Bear," in *Go Down, Moses* (London: Penguin Books, 1960). 6. J. Kilgo, *Deep Enough for Ivorybills* (New York: Anchor Books, Doubleday, 1986). 7. W. Percy, *Love in the Ruins: The Adventures of a Bad Catholic at a Time Near the End of the World* (New York: Dell, 1971). 8. G. Lewbart, *Ivory Hunters* (Malabar, FL: Krieger, 1996). 9. E. Gordon, *Bird Children* (Chicago: P. F. Volland, 1912). 10. A. R. Bird, "Ivory-Bill Is Still King!" *American Forests* 38 (1932): 634–35, 667. 11. D. Eckelberry, "Search for the Rare Ivorybill," in *Discovery. Great Moments in the Lives of Outstanding Naturalists*, ed. J. K. Terres (Philadelphia: J.B. Lippincott, 1961), 195–207. 12. Anonymous, "Death of a Rare Bird," *Audubon Magazine* 58, no. 4 (1956): 162–63. 13. A. Fullingim, "Folklore in the Big Thicket," in *Tales from the Big Thicket*, ed. F. E. Abernathy (Austin: University of Texas Press, 1967), 22–32. 14. A. Y. Gunter, *The Big Thicket. A Challenge for Conservation* (Austin, TX: Jenkins, 1972), 27. 15. R. Hullihan, [Untitled], *Des Moines Register,* June 24, 1979, n.p.; R. Hullihan, "Thief Loses Heart to Rare, Old Bird," *Des Moines Register,* June 1, 1980, n.p.

Epilogue: The Truth Is Out There

1. F. Lewis, *Tales of a Louisiana Duck Hunter* (Franklin, LA: Little Atakapas, 1988). 2. Anonymous, "Status Review on Ivory-Billed Woodpecker," *Endangered Species Technical Bulletin* 10, no. 5 (1985): 7. 3. A. T. Wayne, "Birds of South Carolina," *Contributions from the Charleston Museum* 1 (1910): i–xxi, 1–254. 4. A. A. Allen and P. P. Kellogg, "Recent Observations on the Ivory-Billed Woodpecker," *Auk* 54 (1937): 164–84, esp. 165. 5. R. Ridgway, "The Home of the Ivory-Bill," *Osprey* 3 (1898): 35–36. 6. J. V. Dennis, "A Last Remnant of Ivory-Billed Woodpeckers in Cuba," *Auk* 65 (1948): 497–507. 7. J. T. Tanner, *The Ivory-Billed Woodpecker Research Report No. 1* (New York: National Audubon Society, 1942), 42. 8. D. G. Schueler, *Preserving the Pascagoula* (Jackson: University Press of Mississippi, 1980). 9. J. Dennis, *The Great Cypress Swamps* (Baton Rouge: Louisiana State University Press, 1988), 42. 10. H. Stoddard, *Memoirs of a Naturalist* (Norman: University of Oklahoma Press, 1969), 282. 11. C. H. D. Clarke et al., "Report of the Committee on Bird Protection," *Auk* 81 (1964): 417–25. 12. J. A. Jackson, "The Truth Is Out There," *Birder's World* 16 (2002): 40–47. 13. S. C. Clarke, "Ivory-Billed Woodpecker in Florida," *Forest & Stream* 24 (1885): 367; F. M. Phelps, "The Resident Bird Life of the Big Cypress Swamp Region," *Wilson Bulletin* 26 (1914): 86–101.

Appendix 1: A Geography of Extinction: Where the Ivory-Bills Were

1. T. A. Imhof, *Alabama Birds*, 2nd ed. (University: University of Alabama Press, 1976). 2. P. H. Gosse, *Letters from Alabama* (London: Morgan and Chase, 1850), 91–93; E. M. Hasbrouck, "The

Present Status of the Ivory-Billed Woodpecker (*Campephilus principalis*)," *Auk* 8 (1891): 174–86; W. C. Avery, "Birds Observed in Alabama," *American Field* 34 (1890): 607–08; A. H. Howell, *Birds of Alabama* (Montgomery: Department of Game and Fisheries of Alabama, 1928). 3. Hasbrouck, "The Present Status of the Ivory-Billed." 4. J. J. Audubon, *Journal of John James Audubon Made During His Trip to New Orleans in 1820–1821* (Cambridge, MA: Boston Historical Society, 1929). 5. D. A. James and J. C. Neal, *Arkansas Birds* (Fayettville: University of Arkansas Press, 1986). 6. Ibid. 7. J. A. Jackson, "Ivory-Billed Woodpecker," in *Rare and Endangered Biota of Florida*, ed. J. A. Rodgers, Jr., H. W. Kale II, and H. T. Smith (Gainesville: University Press of Florida, 1996). 8. W. Eastman, "Ten Year Search for the Ivory-Billed Woodpecker," *Atlantic Naturalist* 13 (1958): 216–28; E. S. Austin, *Frank M. Chapman in Florida* (Gainesville: University of Florida Press, 1967). 9. W. E. D. Scott, *The Story of a Bird Lover* (New York: Macmillan, 1903). 10. C. J. Maynard, *The Birds of Eastern North America* (Newtonville, MA: C.J. Maynard, 1881). 11. P. Laurent, "Notes on Birds of Levy County, Florida," *Ornithologist & Oologist* 12 (1987): 157–59. 12. Letter from T. G. Pearson to J. Tanner, September 13, 1940; courtesy of N. Tanner. 13. Austin, *Frank M. Chapman*, 54. 14. C. Bendire, "Life Histories of North American Birds, from the Parrots to the Grackles, with Special Reference to Their Breeding Habits and Eggs," *US National Museum Special Bulletin* 3 (1895): 1–518. 15. A. H. Howell, *Florida Bird Life* (Tallahassee: Florida Department of Game and Fresh Water Fish, in cooperation with Bureau of Biological Survey, US Department of Agriculture, 1932). 16. J. C. Phillips, "An Attempt to List the Extinct and Vanishing Birds of the Western Hemisphere," *Verhandlungen VI Internationalen Ornithologen-Kongresses Kopenhagen* (1926): 512–13. 17. Ibid.; H. H. Bailey, "The Ivory-Billed Woodpecker in Florida," *Oologist* 44 (1927): 18, 20. 18. Tanner, *The Ivory-Billed Woodpecker*. 19. Austin, *Frank M. Chapman*, 111, 127; W. E. Safford, "Natural History of Paradise Key and the Nearby Everglades of Florida," *Annual Report, Smithsonian Institution* 1917 (1919): 377–434; P. Bartsch, "The Bird Rookeries of the Tortugas," *Annual Report Smithsonian Institution* 1917 (1919): 469–500. 20. Howell, *Florida Bird Life*, 313. 21. Jackson, "Ivory-Billed Woodpecker." 22. Tanner, *The Ivory-Billed Woodpecker*; T. D. Burleigh, *Georgia Birds* (Norman: University of Oklahoma Press, 1958); E. E. Murphey, *Observations of the Bird Life of the Middle Savannah Valley, 1890–1937* (Charleston, SC: Charleston Museum, 1937). 23. A. Wilson, *American Ornithology. Vol. 4* (Philadelphia: Bradford and Inskeep, 1811). 24. H. W. Coolidge, "Record of an Ivory Billed Woodpecker Killed Thirty Miles Up River from Savannah, Georgia," *Oriole* 31 (1966): 20–21. 25. Murphey, *Observations of the Bird Life*, 29. 26. R. Brasher, *Secrets of the Friendly Woods* (New York: Century, 1926), 145–46. 27. C. T. Trowell, "The Search for the Ivory-Billed Woodpecker in the Okefenokee Swamp," *OWL News* 6 (1998): 1–7; F. V. Hebard, "Winter Birds of the Okefinokee and Coleraine," *Georgia Society of Naturalists Bulletin* 3 (1941): 1–94. 28. Phillips, *An Attempt to List the Extinct and Vanishing Birds*. 29. A. S. McQueen and H. Mizell, *History of Okefenokee Swamp* (Tallahassee, FL: Rose Printing, 1926). 30. Trowell, "The Search for the Ivory-Billed." 31. F. Harper, "Okefinokee Swamp as a Reservation," *Natural History* 20 (1920): 29–41, esp. 30. 32. A male, Academy of Natural Sciences, Philadelphia, Pennsylvania; ANSP #146,350, Okefenokee Swamp, Georgia, Craven's Island, March 1912. 33. McQueen and Mizell, *History of the Okefenokee*, 140. 34. R. Ridgway, "The Ornithology of Illinois," *Illinois Natural History Survey, Springfield* 1 (1889): 374–76. 35. B. T. Gault, *Checklist of the Birds of Illinois* (Chicago: Illinois Audubon Society, 1923). 36. R. M. Mengel, "The Birds of Kentucky," *Ornithological Monographs* 3 (1965): i–xiv, 1–581. 37. P. W. Parmalee, "Remains of Rare and Extinct Birds from Illinois Indian Sites," *Auk* 75 (1958): 169–76. 38. R. Haymond, "Birds of

Franklin County, Indiana," *Indiana Geological Survey Annual Report* 1 (1869): 209–35; A. W. Butler, "Observations on Faunal Changes," *Bulletin of the Brookville Society of Natural History* 1 (1885): 5–13; A. W. Butler, "List of Birds Observed in Franklin County, Indiana," *Bulletin of the Brookville Society of Natural History* 2 (1886): 12–39; A. W. Butler, *The Birds of Indiana* (Brookville: Indiana Horticultural Society, Amos W. Butler, 1891). 39. A. W. Butler, "The Birds of Indiana," *Indiana Department of Geology and Natural Resources Annual Report* 22 (1897): 515–1187. 40. L. Jones, *The Birds of Ohio* (Columbus: Ohio Academy of Science, 1903). 41. Tanner, *The Ivory-Billed Woodpecker;* R. M. Mengel, "The Birds of Kentucky," *Ornithological Monographs* 3 (1965): i–xiv, 1–581. 42. Mengel, "The Birds of Kentucky." 43. J. J. Audubon, *Journal of James John Audubon,* 33. 44. L. O. Pindar, "Birds of Fulton County, Kentucky," *Wilson Bulletin* 37 (1925): 77–88. 45. A. W. Schorger, "An Early Record and Description of the Ivory-Billed Woodpecker in Kentucky," *Wilson Bulletin* 61 (1949): 235; D. McKinley, "An Early Record of the Ivory-Billed Woodpecker in Kentucky," *Wilson Bulletin* 70 (1958): 380–81. 46. Ibid. 47. H. C. Oberholser, "The Bird Life of Louisiana," *Bulletin of the Louisiana Department of Conservation* 28 (1938): i–xii, 1–834. 48. Tanner, *The Ivory-Billed Woodpecker.* 49. J. J. Audubon and J. B. Chevalier, *The Birds of America. Volume 4* (New York: Dover, 1840–1844/1967), 214. 50. Tanner, *The Ivory-Billed Woodpecker;* E. M. Hasbrouck, "The Present Status of the Ivory-Billed Woodpecker (*Campephilus principalis*)," *Auk* 8 (1891): 174–86; T. D. Burleigh, "Bird Life of the Gulf Coast Region of Mississippi," *Louisiana State University Museum of Zoology, Occasional Papers* 20 (1944): 329–490; J. A. Jackson, "The History of Ivory-Billed Woodpeckers in Mississippi," *Mississippi Kite* 18 (1988): 3–10. 51. J. V. Dennis, *The Great Cypress Swamps* (Baton Rouge: Louisiana State University Press, 1988); W. H. Turcotte and D. L. Watts, *Birds of Mississippi* (Jackson: University Press of Mississippi, 1999); Letter from M. G. Vaiden to J. Bond, May 20, 1963; Academy of Natural Sciences, Philadelphia. 52. J. D. Corrington, "The Winter Birds of the Biloxi, Mississippi Region," *Auk* 39 (1922): 530–56. 53. Jackson, "The History of the Ivory-Billed Woodpecker"; Letter from Vaiden to Bond. 54. W. W. Cooke, "Report on Bird Migration in the Mississippi Valley in the Years 1884 and 1885," *US Department of Agriculture Bulletin, Division of Economic Ornithology* 2 (1888): 127–28; H. Harris, "Birds of the Kansas City Region," *Transactions of the Academy of Sciences, St. Louis* 23 (1919): 213–371; Tanner, *The Ivory-Billed Woodpecker.* 55. Harris, "Birds of the Kansas City Region." 56. O. Widmann, *A Preliminary Catalogue of the Birds of Missouri* (St. Louis, MO: Academy of Sciences, 1907). 57. American Ornithologists' Union, *Check-List of North American Birds,* 4th ed. (Lancaster, PA: American Ornithologists' Union, 1931). 58. Wilson, *American Ornithology.* 59. Hasbrouck, "The Present Status of the Ivory-Bill." 60. E. Coues and H. C. Yarrow, "Notes on the Natural History of Ft. Macon, N.C., and Vicinity," *Proceedings of the Philadelphia Academy of Natural Sciences* 1878 (1878): 21–28. 61. Hasbrouck, "The Present Status of the Ivory-Billed." 62. Jones, "The Birds of Ohio," 226–27. 63. B. G. Peterjohn, *The Birds of Ohio* (Bloomington: Indiana University Press, 1989). 64. G. M. Sutton, *Oklahoma Birds* (Norman: University of Oklahoma Press, 1967), 321. 65. W. W. Cooke, "Migration in the Mississippi Valley," *Ornithologist & Oologist* 9 (1884): 25. 66. W. W. Cooke, "Some Winter Birds of Oklahoma," *Auk* 31 (1914): 473–93. 67. P. Hahn, *Where Is that Vanished Bird?* (Toronto, Ontario, Canada: Royal Ontario Museum, University of Toronto Press, 1963). 68. J. S. Tomer and M. J. Brodhead, eds., *A Naturalist in Indian Territory. The Journals of S. W. Woodhouse, 1849–50* (Norman: University of Oklahoma Press, 1992). 69. Tomer and Brodhead, *A Naturalist in Indian Territory.* 70. Letter from E. Comby to J. Baker, May 28, 1939. 71. A. Sprunt and E. B. Chamberlain, *South Carolina Bird Life* (Columbia: University of

South Carolina Press, 1970). 72. Tanner, *The Ivory-Billed Woodpecker;* and Sprunt and Chamberlain, *South Carolina Bird Life.* 73. W. Hoxie, "Notes on Birds of the Sea Islands," *Ornithologist & Oologist* 10 (1885): 62–63. 74. W. Hoxie, "Probable Occurrence of the Ivory-Billed Woodpecker on Pritchard's Island, South Carolina," *Ornithologist & Oologist* 12 (1887): 122. 75. Murphey, *Observations of the Bird Life.* 76. Ibid. 77. A. T. Wayne, *Birds of South Carolina. Contributions from the Charleston Museum 1* (Charleston, SC: Charleston Museum, 1910). 78. Tanner, *The Ivory-Billed Woodpecker.* 79. G. C. Dickey, "Call of the Ivory-Bill," *South Carolina Wildlife* 21 (1974): 17–21. 80. R. M. Carter, *Finding Birds in South Carolina* (Columbia: University of South Carolina Press, 1993), 200. 81. W. H. Deaderick, "Audubon in Tennessee," *Migrant* 11 (1940): 59–61. 82. H. C. Oberholser and E. B. Kincaid, Jr., *The Bird Life of Texas. Vol. 1* (Austin: University of Texas Press, 1974); J. J. Audubon and J. B. Chevalier, *The Birds of America. Vol. 4* (New York: Dover, 1840–1844/1967). 83. H. E. Dresser, "Notes on the Birds of Southern Texas," *Ibis* (n.s.) 1 (1865): 466–95. 84. R. H. Baker, "Remarks on the Former Distribution of Animals in Eastern Texas," *Texas Journal of Science* 8 (1956): 356–59. 85. Ibid. 86. J. K. Strecker, "The Birds of Texas," *Baylor University Bulletin* 15 (1912): 1–70. 87. Phillips, *An Attempt to List the Extinct and Vanishing Birds.* 88. Tanner, *The Ivory-Billed Woodpecker;* A. A. Allen, "Ivory-Billed Woodpecker," in *Life Histories of North American Woodpeckers,* ed. A. C. Bent (Washington, DC: US National Museum), 1–12. 89. Oberholser and Kincaid, *The Bird Life of Texas.* 90. Ibid. 91. C. Haynie, Specimens not located. Texas Bird Records Committee, Texas Ornithological Society, copied from Internet, June 10, 1996, from http://members.Tripod.com/~tbrc/snl.htm 92. Oberholser and Kincaid, *The Bird Life of Texas.* 93. G. A. Hall, "West Virginia Birds," *Carnegie Museum of Natural History Special Publication* 7 (1983): 1–188. 94. K. W. Haller, [Untitled], *Redstart* 7 (1940): 65–66.

Index

The following index focuses on individuals; common names of woodpeckers, other birds, and animals mentioned with relevance to ivory-bills, and endangered species; states, rivers, swamps, and other places of importance; and organizations. Attributes of the ivory-billed woodpecker are indexed under "woodpecker: ivory-billed." Scientific names are cross-listed with common names that are listed alphabetically in Appendix 2 and are not indexed. James T. Tanner is not included in the index; his work is fundamental and found throughout the book.